CONFINED SPACE ENTRY

HOW TO ORDER THIS BOOK

BY PHONE: 800-233-9936 or 717-291-5609, 8AM–5PM Eastern Time

BY FAX: 717-295-4538

BY MAIL: Order Department
Technomic Publishing Company, Inc.
851 New Holland Avenue, Box 3535
Lancaster, PA 17604, U.S.A.

BY CREDIT CARD: American Express, VISA, MasterCard

BY WWW SITE: http://www.techpub.com

PERMISSION TO PHOTOCOPY–POLICY STATEMENT

CONFINED SPACE ENTRY

A GUIDE TO COMPLIANCE

Frank R. Spellman, CSP

LANCASTER · BASEL

Confined Space Entry
a **TECHNOMIC** publication

Technomic Publishing Company, Inc.
851 New Holland Avenue, Box 3535
Lancaster, Pennsylvania 17604 U.S.A.

Printed in the United States of America
10 9 8 7 6 5 4 3 2 1

Main entry under title:
Confined Space Entry: A Guide to Compliance

A Technomic Publishing Company book
Bibliography: p.
Includes index p. 133

Library of Congress Catalog Card No. 98-86720
ISBN No. 1-56676-704-0

To
Lisa Spangler and Lydia Steinke
and the
Entire Technomic Production Staff

Table of Contents

Preface

MY primary purpose in writing this book was to bring together (in one text) all of the Occupational Safety and Health Administration's (OSHA's) regulatory requirements for making safe and proper confined space entries. Because confined space entry is a complicated procedure—and a process that contains inherent risks—those concerned with safety in the work place are constantly concerned with how to reduce the risks associated with confined space entry, how to eliminate or decrease the hazards workers face in confined spaces, and how to prevent injuries and fatalities from occurring in confined spaces. But comprehensive materials on confined space entry are difficult to find. Surprisingly, little material on the subject is commercially available.

To fill what I see as a real industry need, I have collected all of the associated requirements and regulations, including OSHA's confined space, lockout/tagout, respiratory protection standards and hot work permit requirements in this guidebook. These separate, specific safety standards and requirements have been combined and organized—as they should be, because each is married to the other—in a way that enables you (the user) to easily determine the critical relationship(s) between and among them—but more importantly, to teach you how to enter confined spaces safely—and how to provide workers with effective training for proper confined space entry.

Confined Space Entry: A Guide to Compliance is written for workers in the field and for managers and technical personnel, who should find it a useful guideline and reference in dealing with complex confined space entry procedures. Government officials and regulatory personnel (as well as members of the legal profession) may find this guidebook gives them a clear idea of what confined space entry entails, not just an overview of the subject.

Written in user-friendly, jargon-free plain English, this guidebook provides you with clear sample programs to serve as models when you write your own programs.

Workers have a growing need for more knowledge of the hazards of their work environments, especially of confined spaces. To fulfill this imperative need, individuals and government must work together to better inform—and protect—these workers as they are exposed to a variety of complex and potentially life-threatening situations in confined spaces.

The information presented in this book reflects expertise gained through years of safety experience and hundreds of confined space entries (some that were almost fatal). Over the course of years, the individual safety programs have been scrutinized, amended, updated, and corrected (via advice and guidance) by OSHA auditors who have examined these programs on numerous occasions.

The message this text delivers is simple: The better both workers and management understand the potential hazards and the implementation of measures to either eliminate or reduce the risks and hazards of confined space entry, the safer the workers and the facility—and the better the relationship between the operating facility, the workers, the community, and the regulators.

Often, inherently dangerous operations such as confined space entry are relegated to personnel who lack the practical knowledge required for a thorough understanding of the hazards of confined space entry. Do you doubt this? Don't. If you read a newspaper regularly, anywhere in the world, soon (rather than later) you will read about horrible confined space incidents (with fatalities, usually of the multiple variety) like the one described in the prologue of this guidebook.

As a result of no information, misinformation, no training, no supervision, little or no knowledge, confined space fatalities are real—they occur. They occur far too often.

I hope this book provides you with insight and direction to facilitate safe confined space entry. At the beginning of this preface, I stated my primary purpose for writing this text. Here is another: This guidebook is designed to provide you, the user, with a roadmap, a key—a procedure that will enable you to accomplish confined space entries in accordance with regulatory requirements in the safest manner possible.

Confined spaces can be very unforgiving. Any guidebook on confined space entry should include as a primary goal providing information that will save lives.

FRANK S. SPELLMAN
Virginia Beach, VA

Prologue

CONFINED SPACE—BY ANOTHER NAME, A TOMB

RECENTLY, a construction crew obtained a contract from an industrial complex to clean and preserve the interior surface of a large chemical storage tank. When the construction crew arrived at the work site and peered inside the tank, they quickly realized (even though they were unfamiliar with OSHA's Confined Space Entry Permit Standard) from the horrendous stench emanating from the interior of the tank that it might not be a safe environment in which to work.

The crew foreperson had an idea. She quickly left the tank and her crew and sought out the work site's manager, asking to borrow an air-monitoring device (a sniffer). Reluctantly, the site manager loaned the foreperson a portable air-monitoring device.

Armed with the air monitor, the foreperson went back to the tank and informed her crew that she was going to test the inside atmosphere of the tank for toxic contamination. This is where the first problem arose. The foreperson had never used an air monitor before. She did not know how to use it properly or how to calibrate it.

After about 20 minutes of frustration (she couldn't get the monitor to work), the foreperson gave up trying. Instead, she directed her two crew members to enter the tank to determine what equipment would be needed to start the preservation project. This is where the second problem arose.

Fifteen minutes after the two crew members had entered the tank, the foreperson decided to find out how they were progressing. She walked over to the tank, stuck her head into the entry port, and asked how it was going.

There was no answer.

Suddenly concerned, the foreperson asked again.

Still no answer.

Adapted from Spellman, F. R., *Safe Work Practices*, Technomic Publishing Company, Inc., Lancaster, PA, 1996.

At this point, the foreperson became seriously concerned. In fact, she panicked. And in the great tradition of foolish heroes everywhere—and without thinking clearly—she reacted.

The foreperson grabbed a flashlight and entered the tank. This is where the third problem arose.

What exactly happened next no one knows. One point was very clear, however; the job went terribly wrong.

Later that day, one of the construction officials arrived at the construction site to see how the crew was doing on the tank job. The construction official knew her personnel were on the job, because the company truck was parked next to the tank.

The construction official could not see the crew and decided that they must be working inside the tank. When she looked inside the tank, she saw the dim beam of the flashlight shining on three limp bodies. Obviously, her crew was either taking a nap or . . . something worse. It had to be worse, she realized in horror, because no one could be napping in an environment that smelled that bad.

The construction official feared the worst. She ran into the site office and dialed 911. Later that day, the results were known: three dead workers.

There are lessons to be learned from such a tragedy. Some are obvious. Don't expose your workers (or yourself) to unknown risks. Training is essential for safe confined space entry. Some are not so obvious. For example, no host company official should loan safety equipment to an outside contractor—or to any other non-company person. When you loan your equipment, you may also assume liability for whatever occurs—you buy into the consequences, whether they be good or bad. OSHA came to investigate this fatal incident. Their findings included two major points. First, in this case, loaning plant equipment was the wrong thing to do. When the site official did not show the foreperson how to operate and calibrate the air monitor, he assumed liability for the foreperson's actions—he bought into the incident and its consequences. In the second place, the construction crew had not been trained on proper confined space entry procedures. If they had, this incident might not have occurred at all. With the proper training, that crew would still be alive. The confined space would have been just another confined space, not a tomb.

CHAPTER 1

Introduction

Four men who died Thursday in a large concrete box . . . drowned in a mixture of water and sewage, probably after they were rendered unconscious by some type of gas. . . .

*The concrete box is essentially for holding sewage that is unloaded . . . before the sewage is pumped . . . to sewage lines and eventually to a sewage treatment plant. . . . Sewage commonly generates two types of gases—methane and hydrogen sulfide—that cause loss of consciousness. . . . (*Daily Press*, Newport News-Hampton, VA, p. C10, September 7, 1996)*

1.1 SETTING THE STAGE

UNDER the provisions of the Williams-Steiger Occupational Safety and Health Act (OSH Act) of 1970, employers were assigned the responsibility for occupational and environmental safety, for both workers and the public. The OSH Act, which went into effect on April 28, 1971, directs the U.S. Department of Labor to set forth and enforce safety and health standards for any employer who is involved in a business affecting commerce and who has one or more employees. The term "employer" does not include the United States, any state, or any political subdivision of a state. The term "employee" means anyone who is employed by an employer in a business that affects commerce. When the new act was enacted, the Department of Labor estimated that about 60 million workers were affected by the act.

The act created the position of assistant secretary of labor for occupational safety and health to carry out the responsibilities of the secretary of labor under the act and to head the Occupational Safety and Health Administration (OSHA). OSHA's major responsibilities include enforcing all provisions of the act, setting forth standards and regulations, overseeing state plans, and employer-employee training.

1.1.1 OBJECTIVES OF THE OSH ACT

The principal objective of the OSH Act was to ensure a safe and healthy working environment for all workers. This goal is primarily accomplished by promulgation and enforcement of safety and health standards and regulations. The OSH Act's major objectives include the following:

- to define employer and employee responsibilities and rights in creating a safe and healthy work environment
- to encourage identification and elimination of safety and health problems in the work place
- to keep a record of work-related accidents, fatalities, and injuries
- to encourage states to create their own occupational safety and health programs, which must be at least as effective as the federal program
- to make operations more cost-effective by eliminating industrial accidents and lost-time injuries
- to set forth and enforce safety and health standards and regulations

Although this text recognizes, illustrates, and discusses all of these objectives (in one way or another), the last objective is the focus of this text: to set forth and enforce safety and health standards and regulations. More specifically, we are concerned with the OSHA standards affecting confined space entry, lockout/tagout, respiratory protection, and the hot work permit requirement under the Process Safety Management Standard—all of which are tied into the main topic of this guidebook: confined space entry.

Before we begin a detailed discussion of the standards and requirements pertinent to this text, let's take a closer look at just what the employer and employee responsibilities are. Remember, no matter how many standards and regulations OSHA and other regulators write, promulgate, and attempt to enforce, if employers and employees do not abide by their responsibilities under the act, the requirements are not worth the paper they are written on.

1.1.2 EMPLOYER/EMPLOYEE RESPONSIBILITIES UNDER THE OSH ACT

Section 5 (a) of Public Law 91-596 of December 29, 1970, Occupational Safety and Health Act (OSH Act) requires that

> Each employer—
>
> (1) Shall furnish to each of his employees employment and a place of employment which are free from recognized hazards that are causing or are likely to cause death or serious physical harm to his employees;
>
> (2) Shall comply with occupational safety and health standards promulgated under this Act.

Generally, the employer responsibilities stated above are nothing new to the employer. Doesn't the employer in one way or another (usually based on moral grounds) have a common-sense driven responsibility and obligation to ensure the safety and health of his or her employees? Yes, of course, in the best of all possible worlds this is right—but remember, the reason the OSH Act was enacted (and was needed) in the first place was because employers were not protecting the safety and health of employees on the job. Injury statistics compiled over the years showed an alarming trend toward increasing numbers of on-the-job injuries and lost time. Not only was the frequency of injury occurrence increasing, so was the severity of the injuries. Something had to be done . . . and it was—the result was the OSH Act.

Let's take a look at the employee's responsibilities under the OSH Act.

> Section 5 (b) mandates that
>
> Each employee—
>
> Shall comply with occupational safety and health standards and all rules, regulations, and orders issued pursuant to this Act which are applicable to his own actions and conduct.

This regulation often comes as a surprise to workers, which may surprise you. When workers hear or read the above statement concerning "their" responsibilities under the OSH Act, they are taken aback. Employee perceptions have been that OSHA and its regulations are for the employer and not for the employee. "My boss has to comply and worry about OSHA, not me." I have heard this statement from employees many times in the nearly 30 years since the OSH Act went into effect.

1.2 OSHA STANDARDS

The OSH Act authorizes OSHA to promulgate, modify, or revoke legally enforceable occupational safety and health standards. The standards may require that specific conditions be met or may require the adoption or use of practices, methods, means, or processes that are reasonably necessary and appropriate to protect employees on the job. The employer is required to become familiar with the standards applicable to his or her operation and to make sure that employees have—and use—the personal protective equipment required for safety.

1.2.1 CATEGORIES OF OSHA STANDARDS

The four major categories of OSHA standards are described below.

- *National Consensus or Interim Standards:* These are the safety and health guidelines developed by the American National Standard Institute (ANSI), the National Fire Protection Association (NFPA), and the Fair Labor Standard Act. These guidelines were in existence prior to the OSH Act. When OSHA came into existence, these national consensus standards were adopted and promulgated immediately (The Bureau of National Affairs, 1971).

- *Permanent Standards:* These standards were devised to replace the interim standards. This became necessary when the U.S. Congress realized that occupational safety and health information and knowledge are part of a dynamic (ever-changing) field. Note that in creating new permanent standards, Congress developed procedures to give affected employers and employees a voice in the standard-setting process (The Bureau of National Affairs, 1971).
- *Emergency Temporary Standards:* OSHA has the authority to create and issue temporary or emergency standards any time an employee is in imminent danger. These standards were not meant to take the place of permanent standards, but, instead, their use is intended for limited situations where the health or safety of employees is in grave danger, such as the possibility of exposure to a highly dangerous chemical (The Bureau of National Affairs, 1971).
- *General Standards:* These standards cover hazards common to many industries (four of which concern us in this guidebook). Specifically, we are concerned with the Code of Federal Regulations Title 29, Part 1910 (29 CFR 1910) OSHA General Standards: 1910.146 Confined Space Entry, 1910.147 The Control of Hazardous Energy (Lockout/Tagout), 1910.134 Respiratory Protection, and 29 CFR 119 Process Safety Management (Hot Work Permits).

Together, these standards present a powerful force, dedicated to protecting the lives, health, and well-being of workers. The bottom line: the better both management and workers understand the potential hazards and the implementation of measures to eliminate or reduce the risks and hazards they face on the job, the safer the workers and the facility.

1.3 REFERENCES

Spellman, F. R. *Safe Working Practices for Wastewater Operators.* Lancaster, PA: Technomic Publishing Company, Inc., 1996.

The Bureau of National Affairs. *The Job Safety and Health Act of 1970,* 1st Ed., Bureau of National Affairs, Washington, D.C., 1971.

The Office of the Federal Register. *Code of Federal Regulations Title 29 Parts 1900–1910,* Office of the Federal Register, Washington, D.C., 1995.

CHAPTER 2

Confined Space Entry: The OSHA Standard

Three Newport News Shipbuilding employees are dead after lethal gas and raw sewage flooded a pump room deep within the USS Harry S. Truman, shipyard officials said. . . .

After the leak forced the evacuation of the carrier early Saturday, dozens of emergency crews worked through the night venting flammable gas from the ship so they could safely reach the compartment. Sometime after 1 a.m., rescue workers found the bodies of three pipefitters who were missing and feared dead. (Mark Krewatch, Daily Press, *Newport News-Hampton, VA, p. A1, July 14, 1997)*

2.1 INTRODUCTION

It may surprise you to know (and is alarming to others, including the author) that not until April 15, 1993, did OSHA's Confined Space Standard (29 CFR 1910.146) become effective in all states and territories where OSHA has jurisdiction. States (Puerto Rico and Virgin Islands included) with their own occupational safety and health plans were required to adopt a comparable standard within 6 months.

Why did it take so long to implement a safety standard so critical to protecting the lives and health of workers in industry? A good question—one not easy to answer, especially when you factor in the vagaries of politics and policies (the U.S. military and many shipyards have had such a program for decades). However, that important standard is now in place for general industry. OSHA now has a Confined Space Entry Program.

OSHA's Confined Space Entry Program (CSEP) is a good one—a vital guideline to protect workers and others. CSEP was issued to protect workers who must enter confined spaces. It is designed and intended to protect workers from toxic, explosive, or asphyxiating atmospheres and from possible engulfment from small particles such as sawdust and grain (e.g., wheat, corn, soybean, normally contained in silos). It focuses on areas with immediate health or safety risk—areas with hazards that could potentially cause death or injury—areas of spaces classified as *permit-required* confined spaces. Under the standard, employers are required to identify all permit-required spaces in their work places, prevent unauthorized entry into them, and protect authorized workers from hazards through an entry-by-permit-only program.

CSEP covers all of general industry (Note: this rule does not apply to agriculture 29 CFR 1928, construction 29 CFR 1926, or shipyard employment 29 CFR 1915), including agricultural services (the key word here is "services" and not agriculture), manufacturing, chemical plants, refineries, transportation, utilities, wholesale and retail trade, and miscellaneous services. It applies to manholes, vaults, digestors, contact tanks, basins, clarifiers, boilers, storage vessels, furnaces, railroad tank cars, cooking and processing vessels, tanks, pipelines, silos, among others.

2.2 CONFINED SPACE ENTRY: DEFINITIONS

As with each division of some subject matter that you might study (biology—subdivision: microbiology), most rules, regulations, and standards have their own set of terms essential for communication between managers and the workers required to comply with the guidelines. Therefore, key terms that specifically pertain to OSHA's Confined Space Entry Program are defined and presented here in alphabetical order. The definitions are from OSHA's *Occupational Safety and Health Standards for General Industry (29 CFR 1910 subpart J—General Environment 29 CFR 1910.146 Confined Space Entry,* 1995). The bottom line: Obviously, understanding any rule or regulation is difficult unless you have a clear and concise understanding of the terms used.

2.2.1 DEFINITIONS

- *Acceptable entry conditions* the conditions that must exist in a permit space to allow entry and to ensure that employees involved with a permit-required confined space entry can safely enter into and work within the space.
- *Attendant* an individual stationed outside one or more permit spaces who monitors the authorized entrants and who performs all attendant's duties assigned to the employer's permit space program.
- *Authorized entrant* an employee who is authorized by the employer to enter a permit space.
- *Blanking and blinding* the absolute closure of a pipe, line, or duct

by the fastening of a solid plate (such as a spectacle blind or a skillet blind) that completely covers the bore and that is capable of withstanding the maximum pressure of the pipe, line, or duct with no leakage beyond the plate.

- *Confined space* a space that is large enough and so configured that an employee can bodily enter and perform assigned work; has limited or restricted means for entry or exit (e.g., tanks, vessels, silos, storage bins, hoppers, vaults, and pits are spaces that may have limited means of entry); and is not designed for continuous employee occupancy.
- *Double block and bleed* the closure of a line or pipe by closing and locking or tagging (Note the interface with lockout/tagout) a drain or vent valve in the line between the two closed valves.
- *Emergency* any occurrence or event (including any failure of hazard control or monitoring equipment) internal or external to the permit space that could endanger entrants.
- *Engulfment* the surrounding and effective capture of a person by a liquid or finely divided (flammable) solid substance that can be aspirated to cause death by filling or plugging the respiratory system or that can exert enough force on the body to cause death by strangulation, constriction, or crushing.
- *Entry* the action by which a person passes through an opening into a permit-required confined space. Entry includes ensuing work activities in that space *and is considered to have occurred as soon as any part of the entrant's body breaks the plane of an opening into the space.*

Note 1: The italics in this definition are added for emphasis. Why? Because in the past, many workers thought that the actual entering of a confined space meant that they needed to place their bodies actually into the space—had to actually physically go into the area. Many companies (and eventually OSHA) found that under the old rules on confined space entry, the exact meaning of a confined space "entry" was not clearly spelled out. Instead, some of the old rules and regulations used in various industries were vague and ambiguous on this subject. OSHA, under its new 1993 rule, moved quickly to clear up the possibility of misunderstanding and to enhance the definition of exactly what constitutes a confined space entry.

One of the interesting and potentially life-threatening sidelights of this particular issue is demonstrated quite clearly in Example 2.1.

Note 2: When OSHA initially issued its Final Rule on Confined Space Entry in early 1993, users in the field were confused about several items in the standard. One item of confusion, for example, was the original "point of entry" definition in the preamble, which stated that doorways and other portals through which a person can walk are not to be considered limited means for entry or exit. This was intended to limit the definition of confined spaces to those areas where an employee would be forced to enter or exit in a posture that not only might not be comfortable, but that might also (and more importantly) slow self-rescue or make rescue more difficult.

Safety professionals in the field, however, knew from experience that, even if a door or portal of a space is of sufficient size, obstructions could make entry into or exit from the space difficult. Even though OSHA's intent was that spaces that otherwise meet the definition of confined spaces and that have obstructed entries or exits (even though the portal is a standard size doorway) should be classified as confined spaces. The problem was that OSHA's intent was not clear to the reader, to the supervisor, to other qualified or competent people, or to the potential confined space entrant.

Fortunately, OSHA recognized this ambiguity and changed this statement in its 1994 revision of the preamble to read: ". . . OSHA notes that doorways and other portals through which a person can walk are not to be considered limited means for entry or exit. However, a space containing such a door or portal may still be deemed a confined space if an entrant's ability to escape in an emergency would be hindered."

- *Entry permit (permit)* the written or printed document provided by the employer to allow and control entry into a permit space, and that contains the information shown in an approved Entry Permit (see Figure 5.1).
- *Entry supervisor* the person (such as the employer, foreperson, or crew chief) responsible for determining whether acceptable entry conditions are present at a permit space where entry is planned, for authorizing entry and overseeing entry operations, and for terminating entry as required by the Confined Space Entry Standard.

  Note 1: In practice (in the real world of performing confined space entry operations), common routine often designates the entry supervisor as the "competent" or "qualified" person. So designated in writing, the competent or qualified person is that entry supervisor who has had the appropriate training and experience and possesses the knowledge required to supervise and effect safe, correct confined space entries.

  Note 2: An entry supervisor may also serve as an attendant or as an authorized entrant, as long as that person is trained and equipped as required by the Confined Space Entry Standard for each role he or she plays. Also the duties of entry supervisor may be passed from one individual to another during the course of an entry operation.

- *Hazardous atmosphere* an atmosphere that may expose employees to the risk of death, incapacitation, impairment of ability to self-rescue (i.e., to escape unaided from a permit space), injury, or acute illness from one or more of the following causes:
  (1) Flammable gas, vapor, or mist in excess of 10 percent of its lower explosive or lower flammable limit (LEL or LFL—which basically mean the same thing)
  (2) Airborne combustible dust at a concentration that meets or exceeds its LFL/LEL [Note this concentration may be approximated as a condition in which the dust obscures vision at a distance of 5 feet (1.52 m) or less.]
  (3) Atmospheric oxygen concentration below 19.5 percent or above 23.5 percent
  (4) Atmospheric concentration of any substance for which a dose or a permissible exposure limit (PEL) is published in Subpart G (of the 1910 General Industry Standard), Occupational Health and Environmental Control, or in Subpart Z, Toxic and Hazardous Substances, which could result in employee exposure in excess of its dose or PEL. Note: An atmospheric concentration of any substance that is not capable of causing death, incapacitation, impairment of ability to self-rescue, injury, or acute illness due to its health effects is not covered by this provision.
  (5) Any other atmospheric condition that is immediately dangerous to life and health (IDLH).

  Note: For air contaminants for which OSHA has not determined a dose or permissible exposure limit, other sources of information [Material Safety Data Sheets (MSDS) that comply with the Hazard Communication Standard (commonly known as HazCom), §1910.1200 of the General Industry Standard, published information, and internal documents] can provide guidance in establishing acceptable atmospheric conditions.
- *Hot Work Permit* the employer's written authorization to perform operations (e.g., riveting, welding, cutting, brazing, burning, and heating) capable of providing a source of ignition (hot work permits are covered in Chapter 14).
- *Immediately dangerous to life or health (IDLH)* any condition that poses an immediate or delayed threat to life, that would cause irreversible adverse health effects, or that would interfere with an individual's ability to escape unaided from a permit space.

  Note: Some materials—hydrogen fluoride gas and cadmium vapor, for example—may produce immediate transient effects that, even if severe, may pass without medical attention, but are followed by sudden, possibly fatal collapse 12–72 hours after exposure. The victim "feels normal" from recovery from these transient effects until collapse. Such materials in hazardous quantities are considered to be "immediately" dangerous to life or health.

- *Inerting* the displacement of the atmosphere in a permit space by a non-combustible gas (such as nitrogen) to such an extent that the resulting atmosphere is non-combustible. Note: This procedure produces an IDLH oxygen-deficient atmosphere.
- *Isolation* the process by which a permit space is removed from service and completely protected against the release of energy and material into the space by such means as blanking or blinding; realigning or removing sections of lines, pipes, or ducts; a double block and bleed system; lockout or tagout of all sources of energy; or blocking or disconnecting all mechanical linkages (OSHA's requirements for lockout/tagout are covered in Chapter 12).
- *Line breaking* the intentional opening of a pipe, line, or duct that is or has been carrying flammable, corrosive, or toxic material, an inert gas, or any fluid at a volume, pressure, or temperature capable of causing injury.
- *Non-permit confined space* a confined space that does not contain or (with respect to atmospheric hazards) have the potential to contain any hazard capable of causing death or serious physical harm.
- *Oxygen-deficient atmosphere* an atmosphere containing less than 19.5 percent oxygen by volume.
- *Oxygen-enriched atmosphere* an atmosphere containing more than 23.5 percent oxygen by volume.
- *Permit-required confined space (permit space)* a confined space that has one or more of the following characteristics:
  (1) Contains or has a potential to contain a hazardous atmosphere
  (2) Contains a material that has the potential for engulfing an entrant
  (3) Has a configuration such that an entrant could be trapped or asphyxiated by inwardly converging walls or by a floor that slopes downward and tapers to a smaller cross section
  (4) Contains any other recognized serious safety or health hazard
- *Permit-required confined space program (permit space program)* the employer's overall program for controlling (and where appropriate, for protecting employees from) permit space hazards and for regulating employee entry into permit spaces.
- *Permit system* the employer's written procedure for preparing and issuing permits for entry and for returning the permit space to service following termination of entry.

- *Prohibited condition* any condition in a permit space that is not allowed by the permit during the period when entry is authorized.
- *Rescue service* the personnel designated to rescue employees from permit spaces.
- *Retrieval system* the equipment [including a retrieval line, chest or full-body harness, wristlets (if appropriate) and a lifting device or anchor—usually a tripod and winch assembly] used for non-entry rescue of people from permit spaces.
- *Testing* the process by which the hazards that may confront entrants of a permit space are identified and evaluated. Testing includes specifying the tests that are to be performed in the permit space.

  Note: Testing enables employers both to devise and implement adequate control measures for the protection of authorized entrants and to determine if acceptable entry conditions are present immediately prior to, and during, entry.

Example 2.1

A few years ago, a worker for a large metropolitan utility company in the midwestern United States was killed in a confined space.

This incident evolved around an organization practice whereby utility workers, during their rounds of sewage control stations, had literally been taught (via on-the-job training) to avoid using a confined space permit, the two-person rule, and other safety practices whenever certain vaults had to be entered to adjust one certain valve in each vault. The valves were wall-mounted just inside the vaults and were designed to be manipulated (from inside the vaults) by 8-inch diameter valve wheel handles. These particular valves (in 22 similar vaults) had to be adjusted each day by one of the assigned station checkers.

One thing a safety professional quickly learns about workers is that if there is any easy way to accomplish a certain task, it's a safe bet the workers will find it.

After the valves were initially installed, in only a short time, the workers discovered that they could manipulate the valve wheel (and thus the valve itself) by remaining outside the vault door and reaching inside, stretching their reach, grasping the valve wheel, and turning it. This was a difficult procedure and required a certain amount of dexterity on the worker's part. However, the workers felt that opening or closing the valve from the outside was easier than having to take the time to fill out a confined space permit, get a team together to sample the air within the vault (which routinely registered more than 300 ppm of hydrogen sulfide and above the LEL for methane—an engineering study dealing with these vault's inherent ventilation problems had been in process for several years), don a self-contained breathing apparatus, wear a safety harness attached to a tending line, and then finally enter the space to perform the work at hand. Management encouraged it. Work was accomplished more quickly—and more cheaply—with a single worker adjusting the valves, rather than a confined space team.

You can probably figure out the eventual outcome of this practice. On the day that a worker died in one of these confined spaces, he was in a hurry. He fully understood that the sewage and other waste in the bottom of the well within this vault was not only foul-smelling, but also that it generated hydrogen sulfide and methane gases. He understood this. He also understood that his health, safety, and well-being could be adversely affected if he were not careful. He understood that he could lose his life: hydrogen sulfide in the right concentrations is a killer, and methane gas in the right amount is not only explosive, but also asphyxiating.

He understood this—but he was in a hurry. Besides, for years, he had been working these valves the way he had been taught—why worry? Today was no different than any of the other times he'd done the same task. Why would he want to change a practice that had served him well in the past? Besides, at that moment, he wasn't thinking of any reason why he should change. He was just going through the daily work routine—as quickly as possible.

He opened the rusty metal door, swinging it wide on stiff hinges (quite a task in itself because the door was the original door, installed about 50 years before), and the horrible stench almost overwhelmed him. Outside the vault, in clean, sweet-smelling air, he took a deep breath, closed his mouth and held his breath. He didn't inhale as he (carefully, with his back almost facing inward) reached in with his left hand to the right side of the door toward the wall with the valve wheel. He grasped the valve wheel and tried to turn it clockwise (in the open direction). As usual, this particular valve was difficult to turn. The worker strained hard to break it loose so that it would turn.

It wouldn't budge.

He tried harder.

He stopped, stepped completely outside the vault, exhaled, and breathed in a huge amount of fresh air. The stench that drifted through the open doorway was almost more than he could handle, but handle it he did, at least for the moment. After a short rest, he returned to the task at hand and tried the valve wheel again.

The valve still wouldn't budge.

He leaned inward a bit more to get a better grip on the valve wheel and to put more force into his motion.

The valve still wouldn't move. He leaned inside a few inches more, his lungs bursting, and tried to turn the valve again. It still would not budge, so he tried it again—but his hand slipped and he fell hard to the floor. When he hit the floor, the remaining air in his lungs was forced out by the impact. Just as the impact and reflex forced him to wildly exhale, reflex made him inhale just as wildly, taking in a deep breath of hydrogen sulfide at more than 600 parts per million (ppm) and methane at 20 percent above the lower explosive limit (LEL), a potent deadly brew. During his circumvention of confined space safe work practices, at times, about half his body had been within the confined space. Now, his body was completely inside the vault, where it remained for two more hours before his remains were discovered and removed by another station checking crew (properly equipped with SCBAs, harnesses, and tending lines) dispatched to find the worker.

This example clearly illustrates the reason and the logic behind OSHA's ensuring that the definition of confined "entry" includes clear language defining what constitutes "entry"—that is, when any part of the entrant's body breaks the plane of an opening into the space, an entry has been made and is subject to the confined space entry regulations.

2.3 REFERENCE

The Office of the Federal Register. *Code of Federal Regulations Title 29 Parts 1900—1910 (.146)*, Office of the Federal Register, Washington, D.C., 1995.

CHAPTER 3

Evaluating the Work Place

The Employer shall evaluate the workplace to determine if any spaces are permit-required confined spaces. [(c) (1), The Office of the Federal Register, 1995]

3.1 INTRODUCTION

Do you need to comply with OSHA's Confined Space Entry Standard? It depends—and OSHA wants all of us to make that determination by evaluating our work places.

So, how do we go about evaluating our work places to determine if we must comply? This question is answered in this chapter.

But first, a note of caution. In the evaluation procedure that follow—and that you must follow to evaluate your work place—you must take every care and caution that you do not walk into, climb into, or crawl into any space unless you are absolutely certain that it is safe to do so. In short, for safety, you must assume any unfamiliar confined space presents hazards, until you have determined by examination and testing that it does not.

3.2 THE EVALUATION PROCESS

To determine if a particular work site must comply with OSHA's Confined Space Entry Standard, we must take certain steps. First, we must be familiar with what a confined space is. Recall that, in Chapter 2, we defined a confined space as a space that is large enough and so configured that an employee can bodily enter and perform assigned work; has limited or restricted means for entry or exit (e.g., tanks, vessels, silos, storage bins, hoppers, vaults, and pits are spaces that may have limited means of entry); and is not designed for continuous employee occupancy.

The next step is to survey the plant site, the facility, the factory, or other type of work site to determine if any spaces or structures fall under OSHA's definition of a confined space. While performing such a survey, you must record on paper the name and location of each space or structure identified for evaluation later. You should also have a list of all work-site confined spaces. This list should be distributed to all employees, placed in plain view on employee bulletin boards, and inserted into your site's written confined space program. One thing is certain—when OSHA audits your facility, it will want to see your list of confined spaces.

To facilitate compliance with the work-site evaluation process in identifying any and all confined spaces, OSHA has designed a decision flow chart (see Figure 3.1).

During the evaluation survey process, if confined spaces are identified, then the determination must be made whether or not they are "permit-required" or "non-permit" confined spaces. To do this, you must be familiar with OSHA's definitions for both. Recall the definitions in Section 2.2:

(1) Non-permit confined space is a confined space that does not contain or (with the respect to atmospheric hazards) have the potential to contain any hazard capable of causing death or serious physical harm.
(2) Permit-required confined space (permit space) is a confined space that has one or more of the following characteristics:
    - contains or has a potential to contain a hazardous atmosphere
    - contains a material that has the potential for engulfing an entrant
    - has an internal configuration such that an entrant could be trapped or asphyxiated by inwardly converging walls or by a floor that slopes downward and tapers to a smaller cross section
    - contains any other recognized safety or health hazard

Now that we are clear on the difference between a non-permit and permit-required space, what's next?

If we find, for example, a space that is obviously a permit-required confined space (for any of the reasons stated above), we are then required to label such a space. Figure 3.2 shows a sample label. If you prefer, the label can also be stenciled on the entrance to or near the entrance to a confined space, as long as the label is clearly visible. The point is the permit-required confined space must be clearly labeled to inform employees of the location—and the danger—posed by the permit-required space.

After identifying and labeling all site permit-required confined spaces, the employer has two choices: (1) to designate such spaces as "off limits" to entry by any employee—and to effectively prevent unauthorized entry—or (2) to develop a written confined space program.

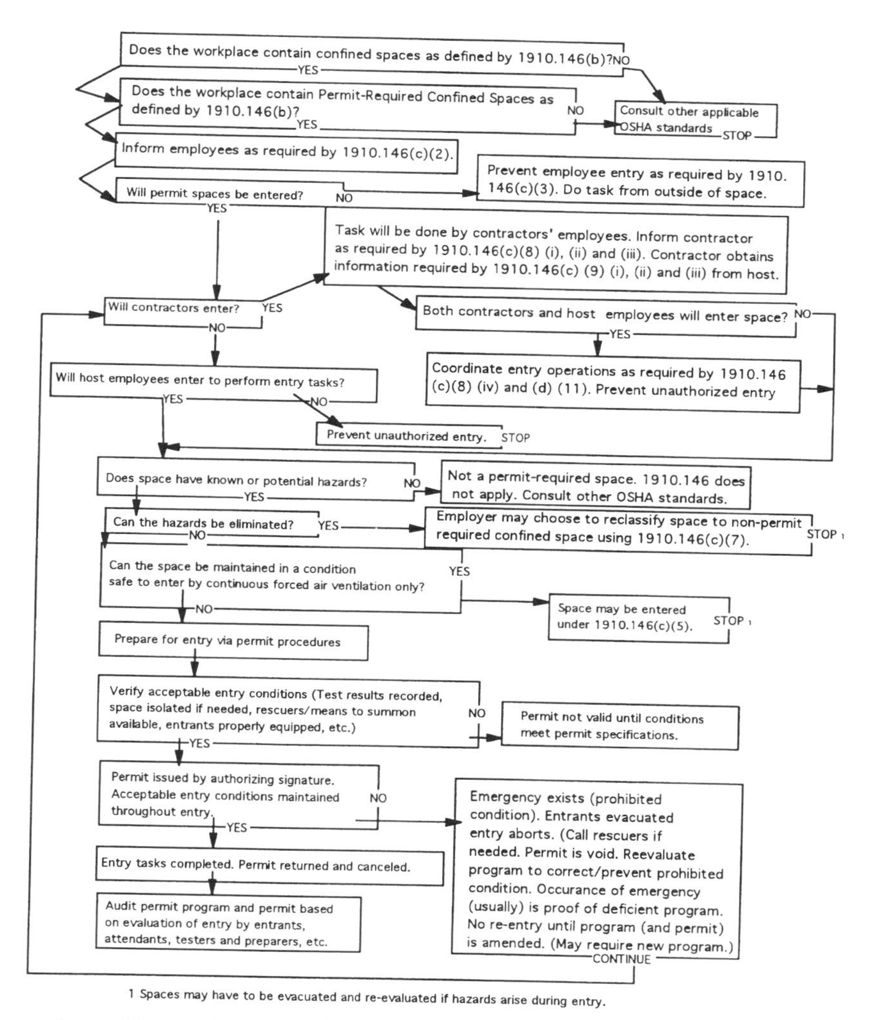

Figure 3.1 Compliance flow chart (*Source:* The Office of the Federal Register, 1995).

Figure 3.2 Sample label.

The requirements for a written confined space program are covered in the next chapter.

3.3 REFERENCE

The Office of the Federal Register. *Code of Federal Regulations Title 29 Parts 1900–1910 (.146),* Office of the Federal Register, Washington, D.C., 1995.

CHAPTER 4

Permit-Required Confined Space Program

If the employer decides that its employees will enter permit spaces, the employer shall develop and implement a written permit space program . . . the written program shall be available for inspection by employees and their authorized representatives. [(c) (4), The Office of the Federal Register, 1995]

4.1 INTRODUCTION

WHEN an employer has identified any work site permit-required confined spaces, he or she can either prohibit the entry of any organizational personnel from entering such spaces or he or she must develop a permit-required confined space program. In this chapter, we assume the employer has identified the existence of permit-required confined spaces on company property and has determined the employees will be required to enter (even if only occasionally) such spaces.

4.2 PERMIT-REQUIRED CONFINED SPACE PROGRAM

The first step the employer must take in implementing a permit-required confined space program is to take the measures necessary to prevent unauthorized entry. Typically, this is accomplished by (first) labeling all confined spaces (see Figure 3.2). The next step is to list all confined spaces and clearly emphasize to employees that the listed spaces are not to be entered by organizational personnel under any circumstances.

I have pointed out again and again that the employer is responsible for identifying, labeling, and listing all site permit-required confined spaces. In addition, the employer must also identify and evaluate the hazards of each confined space.

Once the hazards have been identified and evaluated, the identity and hazard(s) of each site confined space must be listed in the organization's confined space entry program (obviously, it is important that employees are made well aware of all the hazards).

The next step is to develop procedures and practices for those personnel who are required to enter (for any reason) permit-required confined spaces.

The procedures and practices used for permit-required confined space entry must be *in writing* and, at the very least, must include the following:

- specifying acceptable entry conditions
- isolating the permit space
- purging, inerting, flushing, or ventilating the permit space as necessary to protect entrants from external hazards
- providing pedestrian, vehicle, or other barriers as necessary to protect entrants from external hazards
- verifying that conditions in the permit space are acceptable for entry throughout the duration of an authorized entry

Under OSHA's program, the employer must also provide specified equipment to employees involved in confined space entry. The requirements under this specification and the required equipment are covered in the following section.

4.2.1 PERMIT-REQUIRED CONFINED SPACE ENTRY: EQUIPMENT

OSHA in its Confined Space Entry Standard (1910.146), specifies the equipment required to make a safe and "legal" confined space entry into permit-required confined spaces. Note that this equipment must be provided by the employer—at no cost to the employee. The employer is also required not only to procure this equipment at no cost to the employee, but also to maintain the equipment properly. Most importantly, the employer is also required to ensure the employees use the equipment properly. Let's take a look at the type of equipment required for making a safe and legal permit-required confined space entry.

Note: "Equipment" means approved, listed, labeled, or certified as conforming to applicable government or nationally recognized standards, or to applicable scientific principles. It does not mean jury-rigged or "pulled off the wall" devices that might (or might not) be suitable for use by employees. It means that only safe and approved equipment in good condition is to be used—period.

4.2.1.1 Testing and Monitoring Equipment

Numerous makes and models of confined space air monitors (gas detectors or sniffers) are available on the market, and selection should be based on your facility's specific needs. For example, if the permit-required confined space to be entered is a sewer system, then the specific need is a multiple-gas monitor. This type of instrument is best suited for sewer systems, where toxic and combustible gases and oxygen-deficient atmospheres are prevalent.

No matter what type of air monitor is selected for a specific use in a particular confined space, any user must be thoroughly trained on how to effectively use the device. Users must also know the monitor's limitations and how to calibrate the device according to the manufacturer's requirements. Having an approved air monitor is useless if workers are not trained in its operation or proper calibration.

When choosing an air monitor for use in confined space entry, you must ensure that the monitor selected is not only suitable for the type of atmosphere to be entered, but also that it is equipped with audible and visual alarms that can be set, for example, at 19.5 percent or lower for oxygen, and preset for levels of the combustible or toxic gases it is used to detect.

4.2.1.2 Ventilating Equipment

In many cases, you can eliminate, reduce, or modify atmospheric hazards in confined spaces by ventilating—using a special fan or blower to displace the bad air inside a confined space with good air from outside the enclosure. Whatever blower or ventilator type you choose to use, a certain amount of common sense, and a consideration of the depth of the manhole, size of the enclosure, and number of openings available is required. Keep in mind that the blower must be equipped with a vapor-proof, totally enclosed electrical motor, or a non-sparking gas engine. Obviously, the size and configuration of the confined space dictate the size and capacity of the blower to be used. Typically, a blower with a large-diameter flexible hose (elephant trunk) is most effective.

4.2.1.3 Personal Protective Equipment (PPE)

OSHA requires PPE for confined space entries. The entrant must be equipped with the standard PPE required to make a vertical entry into a permit-required confined space (a full-body harness combined with a lanyard or lifeline) and also the PPE required to protect him or her from specific hazards.

For example, an employee who is to enter a manhole is typically equipped with

(1) An approved hard-hat to protect the head
(2) Approved gloves to protect the hands

(3) Approved footwear (safety shoes) to protect the feet
(4) Approved safety eye wear or face protection to protect the eyes and face
(5) Full body clothing (long sleeve shirt and trousers) to protect the trunk and extremities
(6) A tight-fitting National Institute for Occupational Safety and Health (NIOSH) approved self-contained breathing apparatus (SCBA) or supplied air hose-mask with emergency escape bottle—for IDLH atmospheres (respiratory protection is covered in greater detail in Chapter 13).

4.2.1.4 Lighting

Many confined spaces could be described as nothing more than dark (and sometimes foreboding) holes in the ground—often a fitting description. As you might guess, typically, many confined spaces are not equipped with installed lighting. To ensure safe entry into such a space, the entrants must be equipped with intrinsically safe lighting.

Intrinsically safe? Absolutely.

Think about it. The last thing you want to do is to send anyone into a dark space (filled with methane) with a torch in his or her hand—and a light source that emits sparks might as well be a torch. Confined spaces present enough dangers on their own without adding to the hazards. However, even after the space has been properly ventilated (with copious and continuous amounts of outside fresh air) and with the source of methane shut off (blinded or blanked, etc.), we still, obviously, have a space that has the potential for an extremely explosive atmosphere. Do not underestimate the hazards such a confined space presents!

So, what do we do? If lighting is required in a confined space, we need to ensure that it is provided to the entrant for his or her safety, as well as to enable work to be done. For confined space entries, explosion-proof lanterns or flashlights (intrinsically safe devices) are recommended. These devices (if NIOSH and OSHA approved) are equipped with spring-loaded bulbs that, upon breaking, eject themselves from the electrical circuit, preventing ignition of hazardous atmospheres.

Another safe, low-cost, constant light source now readily available for confined space entry is lightsticks. They can be used safely near explosive materials because they contain no source of ignition. Lightsticks are available with illumination times from 30 minutes to 12 hours. The lightsticks shown in Figure 4.1 are activated by simply tossing the lightstick on the ground or against a wall, which breaks the inner glass capsule—illumination is immediate.

Another common work light used for confined space entry is the

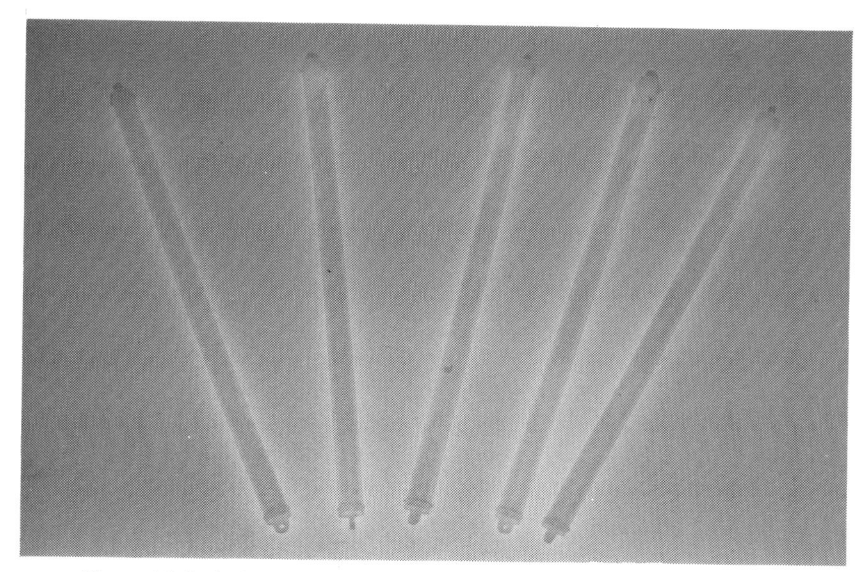

Figure 4.1 Intrinsically safe lightsticks that can be used in confined space entry.

droplight. UL-approved droplights that are vapor-proof, explosion-proof, and equipped with ground fault circuit interrupters (GFCIs) are the recommended type for confined space entry.

Note: If you have a confined space with permanently installed light fixtures that has the potential for an explosive atmosphere in place, the lights must be certified for use in hazardous locations and maintained in excellent condition.

4.2.1.5 Barriers and Shields

As safety professionals, we are concerned not only with the safety of the confined space entrant, but also with the safety of those outside the confined space. For example, an open manhole obviously presents a pedestrian and traffic hazard. To prevent accidents in areas where manhole work is in progress, we can use several safety devices—manhole guard rail assemblies, guard rail tents, barrier tape, fences, and manhole shields, for example. Remember, we not only want to prevent someone from falling into a manhole (or other type of confined space opening) but, also, we want to prevent unauthorized entry. Occasionally, manholes or ordinarily inaccessible areas, when open for work crews, present an attractive nuisance—even ordinary curiosity may lead people (especially children) to put themselves at risk by attempting to enter a confined space.

Along with protecting the confined space opening from someone falling into it or entering it illegally, we must also control traffic around or near the opening. To do this, we may need to employ the use of cones, signs, or stationed guard personnel.

Don't forget the nighttime hours. After dark, it is obviously difficult to see a confined space opening or guard device; these devices should be lighted with vehicle strobes or beacon lights.

4.2.1.6 Ingress and Egress Equipment: Ladders

Have you ever peered inside a 40-foot deep, 24-inch diameter vertical manhole? Not a pleasant sight? Maybe—maybe not. It depends on your point of view. If the manhole has no lighting (as most do not), then you are peering into what appears to be a bottomless pit (and maybe it is). Have you been there? If so, no further explanation is needed. You know that at best, entering any manhole such as the one just described can be a perilous undertaking.

If you have never faced entering a manhole such as the one just described, let's consider an important point. If you are tasked to enter such a confined space, you will obviously be interested in entering it (ingressing) safely (taking all required precautions) and returning safely (egressing).

Experience with assessing safety considerations in confined space areas has shown that many of the installed ladders (in place to allow entry and exit inside confined spaces) are not always in the best material condition. Why? Consider the environment they are constantly exposed to year after year.

Confined spaces may be shrouded in moist, chemical-laden atmospheres—conditions excellent for corroding most metals. Most ladders installed in confined spaces are made of metal. Not only do we require our workers to enter dangerous permit-required confined spaces, but without properly evaluating all of the confined space's conditions, we may also be asking them to enter them in a totally unsafe manner—on equipment that may fail.

Installed ladders within confined spaces must be inspected on a periodic basis to ensure their integrity—their safety. Don't forget about the devices used to hold the ladders in place—the securing or attachment bolts or screws. Most of these are made of metal as well—metal that will corrode and weaken with time. I have found ladders that were literally attached to the wall by rust and rust alone, simply waiting for a victim. Adding weight would send ladder and passenger on a less than thrilling ride—one that would almost certainly result in death. Don't let this happen to you!

How about those spaces that do not have installed ladders? For confined spaces not equipped with ladders, stairways, or some other installed means of ingress and egress, we often employ the use of portable ladders. One way or another, we are required to provide a safe way in and out of a confined space—ladders often fit this need.

Upon occasions (more frequently that we would like), however, ladders or stairways for safe entry or exit are not available, practical, or practicable. When such a situation arises, winches and hoisting devices are commonly used to raise and lower entrants. Remember, any lowering and lifting devices must be OSHA-approved as safe to use. Using a rope attached to the bumper of a vehicle to lower or raise an entrant, for example, is strictly prohibited. Only hand-operated lifting/hoisting devices should be employed. Motorized devices are unforgiving—especially whenever the entrant gets caught up in an obstruction (machinery, pipe, angle iron, etc.) that prevents his or her body from moving. The motorized device doesn't care—it just continues to pull the entrant out (sometimes by body parts only). On a motorized device, a person stuck in a confined space could literally be pulled apart. Don't let such a gruesome event occur on your watch.

OSHA regulations were created to prevent just such gruesome incidents from occurring. But gruesome and fatal events (sometimes involving multiple fatalities) occur. Let's take a look at an actual event (see Example 4.1) that occurred in a confined space—resulting in multiple fatalities.

Example 4.1

April 15, 1989, began as an overcast day in a midwestern city in the United States. When the workday began that morning, a utility work crew of eight people was dispatched from the central office to run and tie-in (hook up) a series of underground fiberoptic cables. The work was to be performed in an underground vault 30 feet below street level.

The crew arrived at the vault about 8:30. After having properly blocked off the street with barricades and traffic cones to route traffic around the manhole entry point, the crew gathered its equipment and waited to be lowered into the underground vault.

The underground vault was a large space that had once served as a pumping station and had later been transformed into a cable vault. Entry into the 24-foot × 36-foot vault could be accomplished either by lifting a 4-inch thick, 12-foot × 12-foot square cement plate in the roadway, or by entering the 24-inch manhole in the center of the cement access plate. In the old days (when the vault contained pumping equipment), work crews would enlist the help of a crane and crew, which would lift the 12-foot × 12-foot plate to allow for easier movement of large equipment and and replacement parts for the pumps in or out of the vault. The crane was also used with a manbasket to safely raise or lower personnel into and out of the vault.

Now that the vault was no longer a pumping station with large equipment within it, lifting the 12-foot × 12-foot plate was rare—most crew chiefs thought there was no longer a need for a crane and manbasket; manhole entry was sufficient.

Besides, the crew chiefs who normally worked these vault entries soon found out that attempting to procure the services of a city crane and crew just to lower and raise personnel into and out of the vault through the manhole was a time-consuming and futile task. The crane and its crew were usually tied up with larger-scale, "more important" work elsewhere. Whenever a crane and its crew were requisitioned to perform cable vault entries, the cable crews would end up standing around, sometimes for hours, waiting for the crane to arrive.

Obviously, the cable crew forepersons didn't like the idea of having to wait on anybody to get their jobs done—so they didn't. They took matters into their own hands. They used the motorized winches attached to the front and rear bumpers of their crew trucks. These winches were used to lower equipment, tools, and workers into and out of the cable vaults—over the years, this had become routine practice.

So, on this particular April morning, it didn't surprise any member of the work crew that the crew chief decided to have the workers enter the vault through the manhole, using the truck's winches—a procedure they had performed hundreds of times before.

During this workday, all went well until time for lunch. At noon, everyone in the crew stopped working and grabbed their lunches. Six members of the crew were within the vault and decided to take their lunch break there—it was easier; they wouldn't have to be lifted out and then lowered back down into the vault. Besides, the crew chief wanted to drive the truck back to the main office during lunch to pick up some needed supplies.

One crew member (an older fellow who was close to retirement) remained topside with a walkie-talkie to stay in contact with the rest of the crew down below. (The crew chief drove the truck and the winch back to the main office—do you see a problem here?)

Lunch was uneventful. At 12:30, the cable crew in the vault went back to their work. Meanwhile, topside, the crew chief returned with his load of supplies and backed the truck up close to the manhole—to make unloading and lowering the supplies via truck winch easier.

When the crew chief had maneuvered the truck into position and set the brake, the topside crew member signaled him to come over to where he was standing. The crew chief got out of the truck (leaving the truck motor running) and walked over to the topside crew member, who handed him the radio—the main office wanted to talk to the chief.

The crew chief had talked to his boss for about 20 minutes on the radio before the conversation was finished. Then, he and the other crew member stood around gabbing for a few minutes and decided to unload the supplies from the truck and lower them into the vault.

They attached the various cable straps and other components that would be needed to complete the work inside the vault and then began lowering the first load. Just before starting the winch to lower the load, the crew chief yelled down into the vault: "Heads up!" There was no answer, but this really didn't mean anything much to the crew chief. He figured that his crew was just busy doing their work.

After the load reached the floor level of the vault, the extent of the tragedy began to unfold.

On the vault floor, the equipment and parts lay there, still secured by the straps. The crew chief yelled down into the vault for someone to disconnect the load from the cable hook so that he could raise the cable and attach another load.

But there was no answer, nor was there any movement.

The crew chief yelled again, then again.

Again, no answer.

"Did they all go to sleep down there?" the crew chief asked the topside crew member.

He yelled down into the vault again.

No answer.

The crew chief hunkered down and peered down into the vault. His range of vision was limited, because of the vault's interior partitions. He couldn't see anyone. Then, he looked up toward the topside crew member, as if to say something, but fell silent, his eyes resting on something that sent a chill throughout his entire body, down to the very depths of his soul—the truck. The truck engine was still running, right over the top of the manhole, the tailpipe dumping a load of death and destruction right into the vault.

Later, investigators found the crew members inside the vault. They were dead, of course; all of them.

When an entrant gets into trouble while inside a confined space, what should we do? When this is the case, OSHA is quite specific on what should and should not be done in rescuing a confined space entrant who is in trouble. OSHA's equipment requirements for effecting a safe confined space rescue is the subject of the next section.

4.2.1.7 Rescue Equipment

When confined space rescue is to be effected by any agency other than the facility itself (emergency rescue service, fire department, etc.), the facility is not required to provide the rescue equipment. However, when confined space rescue is to be performed by facility personnel, proper rescue equipment is required.

Proper rescue equipment? What is it?

Proper rescue equipment basically consists of the equipment needed to remove personnel from confined spaces in a safe manner. "In a safe manner" means "to prevent further injury to the entrants and *any* injury to the rescuers."

Confined space rescue equipment (commonly called retrieval equipment) typically consists of three components: safety harness, rescue and retrieval line, and a means of retrieval.

Let's take a closer look at each of these components.

A full-body harness combined with a lanyard or lifeline evenly distributes the fall-arresting forces among the worker's shoulders, legs, and buttocks, reducing the chance of further internal injuries. A harness also keeps the worker upright and more comfortable while awaiting rescue.

The full-body harness used for confined space rescue should consist of flexible straps that continually flex and give with movement, conforming to the wearer's body—eliminating the need to frequently stop and adjust the harness. Usually constructed of a combination of nylon, polyester, and specially formulated elastomer, the proper harness resists the effects of sun, heat, and moisture to maintain its performance on the job. The full-body harness should include a sliding back D-ring (to attach the retrieval line hook) and a non-slip adjustable chest strap (see Figure 4.2).

The heavy-duty rescue and retrieval line is usually a component of a winch system. Both ends of the retrieval lines should be equipped with approved locking mechanisms of at least the same strength as the lines for attaching to the entrant's harness and anchor point.

The winch systems used today are either an approved two-way system or a three-way system (see Figure 4.3). The two-way system is used for raising and lowering rescue operations whenever a retractable lifeline is not needed. Typical systems feature three independent braking systems; a tough, two-speed gear drive; and approximately 60 feet of steel cable. Three-ways systems offer additional protection when a self-retracting lifeline is used. The winch is usually a heavy-duty model (usually rated at 500 lb or 225 kg) with disc brakes to stop falls within inches and is equipped with a shock-absorption feature to minimize injuries. The proper winch should allow the user to raise and lower loads at an average speed of 10 feet to 32 feet per minute in an emergency.

The means of retrieval usually includes the proper winch with built-in fall protection attached to a 7-foot or 9-foot tripod (see Figure 4.4). The tripod should be of sufficient height to allow the victim to be brought above the rim of the manhole (or other opening) and placed on the ground.

4.2.1.8 Other Equipment

If tools are to be used during a confined space entry or rescue, it may be

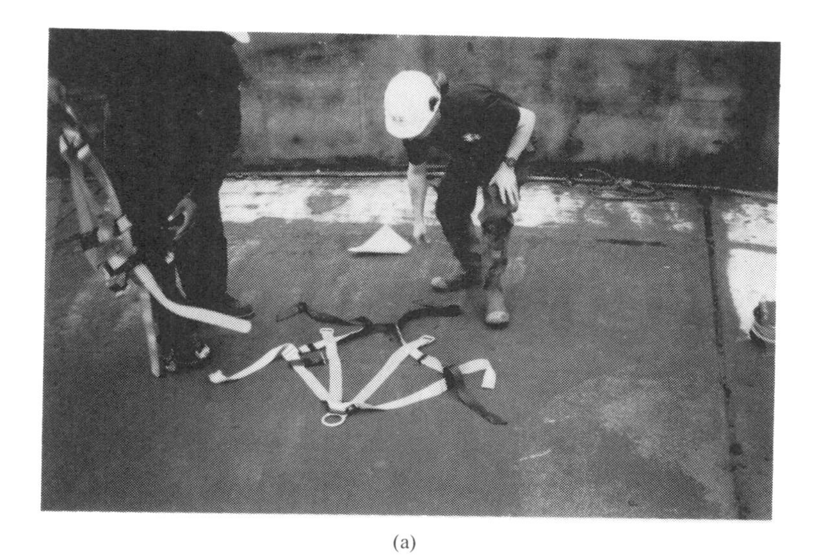

(a)

(b)

Figure 4.2 (a) A full-body harness for confined space rescue. The harness is laid out for inspection. Harnesses should be thoroughly inspected before and after each use. (b) The entrant donning full-body harness. Notice the sliding back D-ring in the helper's right hand.

Figure 4.3 A standard hand-operated winch with safety disk brake system mounted to a tripod.

Figure 4.4 A standard attendant-operated winch, tripod, and attached line to entrant's harness.

necessary to use non-sparking tools if flammable vapors or combustible residues are present. These non-sparking, non-magnetic, and corrosion-resistant tools are usually fashioned from copper or aluminum.

A fire extinguisher, additional radios for communication, spare oxygen bottles (both for SCBAs and cascade systems as needed), a first aid kit, or other equipment necessary for safe entry into and rescue from permit spaces may also be necessary.

4.3 PRE-ENTRY REQUIREMENTS

Before anyone is allowed to enter a permit-required confined space, certain space conditions must first be evaluated. The first step taken should be to determine whether workers must enter the permit-required space to complete the task at hand. You should ask yourself: Do we really need to enter the permit-required confined space? If the answer is yes, then before initiating a confined space entry, the space should be tested with a calibrated air monitor to determine if acceptable entry conditions exist before entry is authorized.

If air monitoring indicates that entry can be made safely without respiratory protection, or if appropriate respiratory protection must be worn, then the supervisor (qualified or competent person) must decide how to effect the entry in the safest manner possible.

Whether the atmosphere is safe or unsafe (without proper respiratory protection), you must ensure that monitoring is continuous. Taking only one reading and basing your decisions on that reading is not wise—in fact, it's unsafe. Conditions can change within a confined space at any time. It is critical to the well-being of the entrant to know when these changes take place and what the changes are.

When conducting the air test for atmospheric hazards, a standard testing protocol should be followed:

- first—test for oxygen
- then—test for combustible gases and vapors
- then—test for toxic gases and vapors

You should also test the atmosphere within a confined space at different levels. For example, if you are about to authorize the entry of workers into a manhole that is 30 feet in depth, you should test, top to bottom, for a stratified atmosphere. Remember, some toxic gases (methane, for example) are lighter than air. They tend to accumulate at the higher levels within the manhole. If the manhole contains carbon monoxide (which has a vapor density similar to air) you should test at the middle level. Hydrogen sulfide (a deadly killer) is heavier than air; therefore, you should test close to the bottom of the manhole. Along with testing at different levels for stratification of toxic gases, you should also check in all directions, to the point possible.

The key point to remember is that atmospheric testing should be continuous, especially when entrants are inside the confined space.

To ensure that continuous atmospheric testing is conducted while an entrant is inside the confined space, an attendant (at least one) must be stationed outside the space to conduct the testing.

In addition to continuously monitoring the atmosphere of the permit-required confined space, the attendant or some other designated person must be familiar with the procedure for summoning rescue and emergency services.

Note: For those facilities having fully trained and equipped on-site rescue teams, it is common practice (and prudent practice) to have the rescue team standing outside the confined space to be on immediate call if required.

Another important function of the attendant or other designated person involved in permit-required confined space entry is to ensure that unauthorized entry into the confined space is prevented.

Before any permit-required confined space entry can be effected, a proper confined space entry permit must be used (see Figure 5.1 later).

When employees from more than one work center (e.g., electricians, machinists, painters, and others from different work centers) or more than one employer are involved in confined space entry, an entry procedure to ensure the safety of all entrants must be developed and implemented.

After the confined space entry is completed, procedures must be in place and used to ensure that the space has been closed off and the permit canceled.

The final step that should be taken after any confined space entry has been effected and is completed is to critique the procedure. Questions should be asked—and answers given. For example: Did anything go wrong during the entry procedure? Did an unauthorized person make an entry into the space? Did any of the equipment used fail? Was anyone injured? Were there any employee complaints about the procedure? Other questions might arise. If questions do come up, steps must be taken to make sure they are answered or that corrections are made to ensure the next entry into a permit-required confined space is a safer one.

At least once each year, the permits accumulated during the year (confined space permits must be retained by the employer for 1 year) should be reviewed. If it is apparent from the review that the procedure should be changed, then change it as needed.

4.4 REFERENCE

The Office of the Federal Register. *Code of Federal Regulations Title 29 Parts 1900–1910 (.146)*, Office of the Federal Register, Washington, D.C., 1995.

CHAPTER 5

Permit System

A permit system for permit-required confined space entry is required by the Confined Space Standard. An entry supervisor (qualified or competent person) must authorize entry, prepare and sign written permits, order corrective measures if necessary, and cancel permits when work is completed. Permits must be available to all permit space entrants at the time of entry and should extend only for the duration of the task. They must be retained for a year to facilitate review of the confined space program. [(1910.146 (e)(f), The Office of the Federal Register, 1995)]

5.1 INTRODUCTION

THE information above sums up OSHA's requirements under its Confined Space Entry Standard (29 CFR 1910.146) and in particular for sections (e) Permits System and (f) Entry Permit.

The gist of OSHA's requirements under these sections follows:

(1) To ensure that a permit is actually used for entry into permit-required confined spaces
(2) To ensure that an entry supervisor (the qualified or competent person) authorizes the entry
(3) To ensure that the entry permit is signed
(4) To ensure that any corrective measures are taken if found necessary
(5) To ensure the permit is canceled when work is completed

Confined space entry permits must be available to all permit space entrants at the time of entry and should extend only for the duration of the task. As stated previously, the permits must be retained for a year to facilitate review of the confined space program.

5.2 PERMIT REQUIREMENTS

What does a confined space permit require, and what does it look like? These are standard questions that may arise any time confined space training is being conducted and at those times when a facility is developing a permit-required confined space program for use and for compliance with OSHA.

Figure 5.1 illustrates a sample confined space permit. This particular permit is included here because it has been successfully used for several years and has been revised as required; it has been tested and has stood the scrutiny of OSHA and insurance auditors. Are there better permits out there? Maybe there's one that would suit your particular requirements more closely. However, if you need assistance in developing your own site-specific permit, the sample shown in Figure 5.1 can assist you in this effort. Additionally, OSHA, in its 1910 Standard, has published sample permits (listed in Appendix D to §1910.146). These samples can also aid you in fashioning your own permit.

What is required to be on a confined space permit?

According to OSHA, an entry permit must include the following (you might want to check Figure 5.1 to see if the requirements are listed as they should be):

(1) Identification of the permit space to be entered
(2) The purpose of the entry
(3) The date and authorized duration of the entry permit
(4) The authorized entrants within the permit space by name, or by such other means as will enable the attendant to determine quickly and accurately, for the duration of the permit, which authorized entrants are inside the permit space
(5) The personnel, by name, currently serving as attendants
(6) The individual, by name, currently serving as the entry supervisor (qualified or competent person), with a space for the signature or initials of the entry supervisor who originally authorized entry
(7) The hazards of the permit space to be entered
(8) The measures used to isolate the permit space and to eliminate or control permit space hazards before entry (what this really means is that lockout/tagout must be completed)
(9) The acceptable entry conditions
(10) The results of initial and periodic tests performed, accompanied by the names or initials of the testers and by an indication of when the tests were performed
(11) The rescue and emergency services that can be summoned and the means (such as the equipment to use and the numbers to call) for summoning those services

CONFINED SPACE PERMIT

Date and Time: ____
Issued   Expires (12 hours max.)

Job Site/Space ID: ____________________ **Job Supervisor:** ____________________

Equipment to be worked on and work to be performed: ____________________

Attendant (s): ____________________

Entrant (s): ____________________

1. Atmospheric Checks:
- Time ______
- Oxygen ______%
- LEL ______%
- H_2S ______ppm
- Tester's Signature: ____________________

2. Source Isolation (No Entry)

	N/A	YES	NO
Pumps or lines blinded, disconnected, blocked, and locked/tagged out (see attachment)	☐	☐	☐

3. Ventilation Modification:

	N/A	YES	NO
Mechanical	☐	☐	☐
Natural Ventilation Only	☐	☐	☐

4. Atmospheric Check After Isolation and Ventilation:
- Oxygen ______% (19.5% - 23%)
- LEL ______% (< 10%)
- H_2S ______ppm (< 10 ppm)
- Time ______
- Tester's Signature: ____________________

5. Entrant, attendant, and rescue persons:

	YES	NO
Successfully completed required training?	☐	☐
Is it current?	☐	☐

6. Equipment:

	N/A	YES	NO
Direct reading gas monitor tested	☐	☐	☐
Safety harnesses and lifelines for entrant and attendants	☐	☐	☐
Hoisting Equipment	☐	☐	☐
Powered Communications	☐	☐	☐
SCBA for entrant and attendants (s)	☐	☐	☐
Protective Clothing	☐	☐	☐
All electrical equipment listed Class I, Division I, Group D and Non-Sparking tools	☐	☐	☐

7. Communication Procedures: ____________________

8. Rescue Procedures: ____________________

9. Hazards In Space: ____________________

10. Hazards Taken Into Space: ____________________

11. Periodic Atmospheric Tests:

	Initial	2	3	4	5	6	7	8	9	10	11
Oxygen (%)	___	___	___	___	___	___	___	___	___	___	___
LEL (%)	___	___	___	___	___	___	___	___	___	___	___
H_2S (ppm)	___	___	___	___	___	___	___	___	___	___	___

We have reviewed the work authorized by this permit and the information contained here-in. Written instructions and safety procedures have been received and are understood. Entry cannot be approved if any squares are marked in the "NO" column. This permit is not valid unless all appropriate items are completed.

Permit Prepared By: (Qualified Person) ____________________
Printed Name   Signature

Approved By: (Supervisor) ____________________
Printed Name   Signature

Reviewed By: (Safety Division Personnel): ____________________
Printed Name   Signature

Figure 5.1 Confined space permit.

(12) The communication procedures used by authorized entrants and attendants to maintain contact during the entry
(13) Equipment, such as personal protective equipment, testing equipment, communications equipment, alarm systems, and rescue equipment
(14) Any other information whose inclusion is necessary, given the circumstances of the particular confined space, in order to ensure employee safety
(15) Any additional permits, such as for hot work, that have been issued to authorize work in the permit space

Remember, with confined space permits, as with any documentation OSHA requires, you must be sure to maintain your records carefully. Any documents you create and use are subject to OSHA auditing—and may at some point be required as evidence in a legal case.

5.3 REFERENCE

The Office of the Federal Register. *Code of Federal Regulations Title 29 Parts 1900–1910 (.146),* Office of the Federal Register, Washington, D.C., 1995.

CHAPTER 6

Confined Space Training

The employer shall provide training so that all employees whose work is regulated by this [standard] acquire the understanding, knowledge, and skills necessary for the safe performance of the duties assigned. . . . [1910.146 (g), The Office of the Federal Register, 1995)]

6.1 INTRODUCTION

ANY work requirement is easier to perform if the person doing the task is fully trained on the proper way to accomplish it. Training offers another advantage as well—increased safety. In accomplishing any work task safely, proper training is critical.

It doesn't take a lot of thought to understand the importance of training to maintain safety in the work place. Though this is the case, you might be surprised to know that many post-accident investigations of on-the-job injuries have indicated that workers were often injured because they lacked the knowledge to correctly and safely perform their assigned tasks.

For example, in one case investigated several years ago, an employee was injured when a basket-lifting device flipped over the top of the machine that was lifting it [a modified (jury-rigged) front-end loader outfitting with forks] with a worker inside the basket. The worker broke every rib along the right side of his body and, as you can imagine, had severe internal injuries. He survived this incident only by chance.

The investigation indicated that the lifting basket was designed to be lifted only by a forklift. The manufacturer's technical manual made this precaution plain, both with words and pictures.

When asked about the manufacturer's safety recommendations, the person in charge of the operation stated that he had seen it, but hadn't had the time to read it.

Obviously, when the supervisor hadn't read the manufacturer's recommended safety requirements for the safe operation of the lifting basket, neither had anyone else—especially the victim.

This incident and the several thousand others like it that occur each year point to the importance of providing safety training to both workers and supervisors.

Confined space entry operations are extremely dangerous undertakings. I stated earlier that confined spaces are very unforgiving—this is the case even for those workers who have been well trained. However, training helps to reduce the severity of any incident. When something goes wrong (as is often the case), it is better to have fully trained personnel standing by than to have people standing by who are not trained—who do not know how to properly rescue an entrant, let alone how to rescue themselves. When you get right down to it, having fully trained workers, for any job, just makes good common sense.

6.2 TRAINING REQUIREMENTS FOR CONFINED SPACE ENTRY

OSHA is very clear on its requirement to train confined space entry personnel. Both initial and refresher training must be provided. This training must provide employees with the necessary understanding, skills, and knowledge to perform confined space entry safely. Refresher training must be provided and conducted whenever an employee's duties change, when hazards in the confined space change, or whenever an evaluation of the confined space entry program identifies inadequacies in the employee's knowledge. The training must establish employee proficiency in the duties required and shall introduce new or revised procedures as necessary for compliance with the standard.

OSHA also requires the employer to certify *in writing* that the employee has been trained. This certification must include the employee's name, the signature of the trainer, and the dates of training. Typically, employers certify this training by conducting written and practical examinations (including training dry runs or drills). When an employee meets the certification requirements, the employee is normally awarded a certificate stating that he or she as been trained and certified (by whatever means). These written certifications should be filed in the employee's personnel and training records.

Any time you conduct safety training, you must keep accurate records of the training. OSHA will want to see these records any time it audits your facility (for whatever reason). Any supervisor or training official who provides critically important (possibly life-saving) training would be foolish not to keep and maintain accurate training records—they may be needed in a legal action.

ATTENDANCE ROSTER

TRAINER: DATE:

CONFINED SPACE TRAINING

In accordance with the recordkeeping and training requirements of the Confined Space Entry Standard, I have received training on Confined Space Entry Procedures. I have agreed to verify my understanding and training on 29 CFR 1910.146 OSHA's Confined Space Entry Standard by signing this roster. This training meets the requirements as specified by 29 CFR 1910.146.

Name:	Work Center:
________________	____________
________________	____________
________________	____________
________________	____________
________________	____________

Figure 6.1 Typical training roster form.

To facilitate the recordkeeping process, a form or roster with a statement like the one shown in Figure 6.1 is highly recommended.

Remember, not only does OSHA require training on its Confined Space Standard and other associated standards (i.e., lockout/tagout, respiratory protection, and hot work permits), this training is critically important to the well-being of workers. By making sure they know that their work organization is taking all possible steps to ensure their safety, they should buy in to the required safe work practices themselves.

You must be able to demonstrate that this training was actually conducted. The form in Figure 6.1 will aid in this effort. Remember, you can do all the safety training you want. The safety training can be of the highest quality possible. You can conduct this training more frequently than OSHA requires. However, one thing is certain, if you do all of these things and do not document the training, in the eyes of OSHA and the legal system, the training was never accomplished. All training must be documented.

6.2.1 THE WORK PLACE CONFINED SPACE TRAINING PROGRAM

Are you at a loss as to what the actual training program should entail for the worker? Exactly what should you include in your work place confined space training program? It depends. Any work place training program on just about any OSHA requirement is somewhat site-specific. For example, confined space training for wastewater workers might be different from the training given to telephone repair people who have to enter underground vaults, because the hazards might not be the same. For example, the underground vault in Example 4.1 presented different hazards when it was used in the sewage system than it does as a cable vault.

As a rule of thumb, it is hard to go wrong on any OSHA-required training if the requirements spelled out in the applicable standard are explained to all workers involved. In addition, for confined space entry training, it is important, at a minimum, to cover the following:

(1) Explain and point out the requirements of 29 CFR 1910.146 (OSHA's Confined Space Standard).
(2) Clearly explain who is responsible for what under the program.
(3) Explain key definitions.
(4) Inform each trainee of the exact location of the work site's permit-required confined spaces.
(5) Explain how to use the work site's confined space permit.
(6) Explain the potential for engulfment.
(7) Explain and demonstrate how to use air-monitoring equipment.
(8) Explain and demonstrate how to use required confined space entry equipment.
(9) Explain the potential for hazardous atmospheres.
(10) Explain the work site's procedures for confined space rescue.
(11) Explain the interface between confined space entry and lockout/tagout, respiratory protection, and hot work permits.
(12) Explain how to properly use the work site's pre-entry checklist.

6.2.2 CONFINED SPACE WRITTEN EXAM (A SAMPLE)

We stated earlier that measuring the employee's level of knowledge of confined space entry procedures is important. One way to accomplish this is to administer a written proficiency examination, such as the sample exam that follows.

Note: You may want to look at and analyze the questions (especially the types of questions) asked in this examination. A prudent course of action on your part would be to ensure that your confined space training program provides the information necessary to enable the workers to answer all these questions.

Permit-Required Confined Space Test[1]

1. What is one of the first questions that should be answered before planning entry into a permit-required confined space?

   Answer: Can this job/task be accomplished without entering the permit space?

2. Confined space hazards are addressed by OSHA in two specific *comprehensive* standards. One of the standards covers general industry and the other covers
   A. Agriculture
   B. Longshoring
   C. Construction
   D. Shipyards

   Answer: D. Shipyards

3. OSHA'S definition of confined spaces in general industry includes
   A. The space being more than 4 feet deep
   B. Limited or restricted means for entry and exit
   C. The space being designed for short-term occupancy
   D. Having only natural ventilation

   Answer: B. Limited or restricted means for entry and exit

4. Which of the following would *not* constitute a hazardous atmosphere under the permit-required confined space standard:
   A. Less than 19.5 percent oxygen
   B. More than the IDLH of hydrogen sulfide
   C. Enough combustible dust that obscures vision at a distance of 5 feet
   D. 5 percent of LEL

   Answer: D. 5 percent of LEL

5. OSHA's review of accident data indicates that the most confined space deaths and injuries are caused by the following three hazards:
   A. Electrical, falls, toxics
   B. Asphyxiants, flammables, toxics

[1]The confined space examination presented here is an adaptation from the examination used by the U.S. Department of Labor—Occupational Health Administration Office of Training and Education—OSHA Training Institute, Des Plaines, Illinois.

C. Drowning, flammables, entrapment
D. Asphyxiants, explosions, engulfment

*Answer: B. Asphyxiants, flammables, toxics (*Federal Register, *1-14-93, p. 4465, 3rd column under "1. Atmospheric Hazards")*

6. Toxic gases in confined spaces can result from
   A. Products stored in the space and the manufacturing processes
   B. Work being performed inside the space or in adjacent areas
   C. Desorption from porous walls and decomposing organic matter
   D. All of the above

   Answer: D. All of the above

7. Oxygen deficiency in confined spaces does *not* occur by
   A. Consumption by chemical reactions and combustion
   B. Absorption by porous surfaces such as activated charcoal
   C. Leakage around valves, fittings, couplings, and hoses of oxy-fuel gas welding equipment
   D. Displacement by other gases

   Answer: C. Leakage around valves, fittings, couplings, and hoses of oxy-fuel gas welding equipment

8. What reading (in percent O_2) would you expect to see on an oxygen meter after an influx of 10 percent nitrogen into a permit space?
   A. 5.0 percent
   B. 11.1 percent
   C. 18.9 percent
   D. 90.0 percent

   Answer: C. 18.9 percent, 100 percent air – 10 percent nitrogen = 90 percent air, 90 percent air × 0.21 percent O_2 = 18.9 percent O_2

9. An attendant is which of the following?
   A. A person who makes a food run to the local 7-11 store for refreshments for the crew inside the confined space
   B. A person who often enters a confined space while other personnel are within the same space
   C. A person who watches over a confined space while other employees are in it and only leaves if he or she must use the restroom
   D. A person with no other duties assigned other than to remain immediately outside the entrance to the confined space and who may render assistance as needed to personnel inside the space. The attendant never enters the confined space and never leaves the space unattended while personnel are within the space.

Answer: D. A person with no other duties assigned other than to remain immediately outside the entrance to the confined space and who may render assistance as needed to personnel inside the space. The attendant never enters the confined space and never leaves the space unattended while personnel are within the space.

10. Per 1910.146, an atmosphere that contains a substance at a concentration exceeding a permissible exposure limit intended solely to prevent long-term (chronic) adverse health effects is *not* considered to be a hazardous atmosphere on that basis alone.
    A. True
    B. False

    *Answer: A. True (*Federal Register, *1-14-93, p. 4474, top of 3rd column)*

11. Of the following chemical substances, which one is a simple asphyxiant and is flammable:
    A. Carbon monoxide (CO)
    B. Methane (CH_4)
    C. Hydrogen sulfide (H_2S)
    D. Carbon dioxide (CO_2)

    Answer: B. Methane (CH_4)

12. Entry into a permit-required confined space is considered to have occurred
    A. When an entrant reaches into a space too small to enter
    B. As soon as any part of the body breaks the plane of an opening into the space
    C. Only when there is clear intent to fully enter the space (therefore, reaching into a permit space would not be considered entry)
    D. When the entrant says, "I'm going in now"

    Answer: B. As soon as any part of the body breaks the plane of an opening into the space

13. If the LEL of a flammable vapor is 1 percent by volume, how many parts per million is 10 percent of the LEL?
    A. 10 ppm
    B. 100 ppm
    C. 1,000 ppm
    D. 10,000 ppm

    Answer: C. 1,000 ppm, 1 percent = 10,000 ppm, 0 percent = 1,000 ppm

14. The principal of operation of most combustible gas meters used for permit entry testing is
    A. Electric arc
    B. Double displacement
    C. Electrochemical
    D. Catalytic combustion

    Answer: D. Catalytic combustion

15. The LEL for methane is 5 percent by volume, and the UEL (upper explosive limit) is 15 percent by volume. What reading should you get on a combustible gas meter when you calibrate with a mixture of 2 percent by volume methane with a balance of nitrogen?
    A. 10,000 ppm (1 percent LEL)
    B. 40 percent LEL
    C. Zero
    D. 80 percent of the flash point

    Answer: C. Zero (Note: If balance had been air: percent volume divided by percent LEL – 2/5 = 40 percent LEL)

16. The proper testing sequence for confined spaces is the following:
    A. Toxics, flammables, oxygen
    B. Oxygen, flammables, toxics
    C. Oxygen, toxics, flammables
    D. Flammables, toxics, oxygen

    Answer: B. Oxygen, flammables, toxics

17. Circle the following true statement(s).
    A. Employers must document that they have evaluated their work place to determine if any spaces are permit-required confined spaces.
    B. If employers decide that their employees will enter permit spaces, they shall develop and implement a written permit space program.
    C. Employers do not have to comply with any of 1910.146 if they have identified the permit spaces and have told their employees not to enter those spaces.
    D. The employer must identify permit-required confined spaces by posting danger signs.

    Answer: B. If employers decide that their employees will enter permit spaces, they shall develop and implement a written permit space program. (Note: See CPL 2.100, p. 18, Question #2 to rule out "A")

18. Circle the following true statement(s).
    A. Under paragraph (c) (5) (i.e., alternate procedures), continuous monitoring can be used in lieu of continuous forced air ventilation if no hazardous atmosphere is detected.
    B. Continuous forced air ventilation eliminates atmospheric hazards.
    C. Continuous atmospheric monitoring is required if employees are entering permit spaces using alternate procedures under paragraph (c) (5).
    D. Periodic atmospheric monitoring is required when making entries using alternate procedures under paragraph (c) (5).

    Answer: D. Periodic atmospheric monitoring is required when making entries using alternate procedures under paragraph (c) (5).

19. OSHA's position allows employers the option of making a space eligible for the application of alternate procedures for entering permit spaces, paragraph (c) (5), by first temporarily "eliminating" all non-atmospheric hazards, then controlling atmospheric hazards by continuous forced air ventilation.
    A. True
    B. False

    Answer: B. False

20. Respirators allowed for entry into and escape from immediately dangerous to life or health (IDLH) atmospheres are
    A. Airline
    B. Self-contained breathing apparatus (SCBA)
    C. Gas mask
    D. Air purifying
    E. A and B

    Answer: B. Self-contained breathing apparatus (SCBA) (Note: A "combination airline with auxiliary SCBA" would be approved, but not an airline.)

21. Circle the following false statement(s).
    A. If all hazards within a permit space are eliminated without entry into the space, the permit space may be reclassified as a non-permit confined space under paragraph (c) (7).
    B. Minimizing the amount of regulation that applies to spaces whose hazards have been eliminated encourages employers to actually remove all hazards from permit spaces.
    C. A certification containing only the date, location of the space, and

the signature of the person making the determination that all hazards have been eliminated shall be made available to each employee entering a space that has been reclassified under paragraph (c) (7).

D. An example of eliminating an engulfment hazard is requiring an entrant to wear a full body harness attached directly to a retrieval system

Answer: D. An example of eliminating an engulfment hazard is requiring an entrant to wear a full body harness attached directly to a retrieval system.

22. Circle the following false statement(s).
    A. Compliance with OSHA's Lockout/Tagout Standard is considered to eliminate electromechanical hazards.
    B. Compliance with the requirements of the Lockout/Tagout Standard is not considered to eliminate hazards created by flowable materials such as steam, natural gas, and other substances that can cause hazardous atmospheres or engulfment hazards in a confined space.
    C. Techniques used in isolation are blanking, blinding, misaligning, or removing sections of line soil pipes and a double block and bleed system.
    D. Water is considered to be an atmospheric hazard.

Answer: D. Water is considered to be atmospheric hazard. (Note: See CPL 2.100, p. 18, Q #11 & Q #12 for additional discussion about water in a confined space.)

23. Circle the following true statement(s).
    A. "Alarm only" devices that do not provide numerical readings are considered acceptable direct reading instruments for initial (pre-entry) or periodic (assurance) testing.
    B. Continuous atmospheric testing must be conducted during permit-required space entry.
    C. Under alternate procedures, OSHA will accept a minimal "safe for entry" level as 50 percent of the level of flammable or toxic substances that would constitute a hazardous atmosphere.
    D. The results of air sampling required by 1910.146, which show the composition of an atmosphere to which an employee is exposed are *not* exposure records under 1910.1020.

Answer: C. Under alternate procedures, OSHA will accept a minimal "safe for entry" level as 50 percent of the level of flammable or toxic substances that would constitute a hazardous atmosphere. (Note: See

CPL 2.100, pp. 19–20, Question #6. To rule out "A," see CPL 2.100, p. 22, Q #16 and to rule out "D," CPL 2.100, p. 24, Q #3.)

24. Example(s) of simple asphyxiants are
    A. Nitrogen (N_2)
    B. Carbon monoxide (CO)
    C. Carbon dioxide (CO_2)
    D. A and C

    Answer: D. A and C (Note: If some workers choose only "A" because CO_2 has a PEL, give them credit.)

25. Which statement(s) is/are true about combustible gas meters (CGMs)?
    A. CGMs can measure all types of gases.
    B. The percent of oxygen will affect the operation of CGMs.
    C. Most CGMs can measure only pure gases.
    D. CGMs will indicate the lower explosive limit for explosive dusts.

    Answer: B. The percent of oxygen will affect the operation of CGMs.

26. Circle the following true statement(s).
    A. An off-site rescue service has to have a permit space program before performing confined space rescues.
    B. The only respirator that a rescuer can wear into an IDLH atmosphere is a self-contained breathing apparatus.
    C. Only members of in-house rescue teams shall practice making permit space rescues at least once every 12 months.
    D. Each member of the rescue team shall be trained in basic first aid and CPR.
    E. To facilitate non-entry rescue, with no exceptions, retrieval systems shall be used whenever an authorized entrant enters a permit space.

    Answer: D. Each member of the rescue team shall be trained in basic first aid and CPR. [See .146 (k)(1)(iv) for correct answer. See CPL 2.100, p. 24, Q #1 under section (k) to rule out "A"; p. 25, Q #3 to rule out "B"; FR 1-14-93, p. 4527, 3rd column, 2nd paragraph to rule out "C"; and see 1146 (k)(3) to rule out "E."]

27. The permit-required confined space standard requires the employer to initially
    A. Train employees to recognize confined spaces
    B. Measure the levels of air contaminants in all confined spaces
    C. Evaluate the work place to determine if there are any confined spaces

D. Develop an effective confined space program

Answer: C. Evaluate the work place to determine if there are any confined spaces

28. If an employer decides that he/she will contract out all confined space work, then the employer
    A. Has no further requirement under the standard
    B. Must label all spaces with a "keep out" sign
    C. Must train workers on how to rescue people from confined spaces
    D. Must effectively prevent all employees from entering confined spaces

Answer: D. Must effectively prevent all employees from entering confined spaces

29. Not required on a permit for confined space entry is
    A. Names of all entrants
    B. Name(s) of entry supervisor(s)
    C. The date of entry
    D. The ventilation requirements of the space

Answer: D. The ventilation requirements of the space

30. Circle the following training requirement that is identical for entrant, attendant, and entry supervisor. Know
    A. The hazards that may be faced during entry
    B. The means of summoning rescue personnel
    C. The schematic of the space to ensure all can get around in the space
    D. The proper procedure for putting on and using a self-contained breathing apparatus

Answer: A. The hazards that may be faced during entry

31. Attendants can
    A. Perform other activities when the entrant is on break inside the confined space
    B. Summon rescue services as long as he/she does not exceed a 200-foot radius around the confined space
    C. Enter the space to rescue a worker but only when wearing a SCBA and connected to a lifeline
    D. Order evacuation if a prohibited condition occurs

Answer: D. Order evacuation if a prohibited condition occurs

32. An oxygen enriched atmosphere is considered by 1910.146 to be
    A. Greater than 22 percent oxygen
    B. Greater than 23.5 percent oxygen
    C. Greater than 20.9 percent oxygen
    D. Greater than 25 percent oxygen when the nitrogen concentration is greater than 75 percent

    Answer: B. Greater than 23.5 percent oxygen

33. The following confined space that would be permit-required is
    A. A grain silo with inward sloping walls
    B. A 10-gallon methylene chlorine reactor vessel
    C. An overhead crane cab that moves over a steel blast furnace
    D. All of the above

    Answer: A. A grain silo with inwardly sloping walls

34. A written permit space program requires
    A. That the employer purchase SCBAs and lifelines, but the employees purchase safety shoes and corrective lens safety glasses
    B. That the employer test all permit-required confined spaces at least once per year or before entry, whichever is most stringent
    C. That the employer provide one attendant for each entrant up to five and one for each two entrants when there are more than five
    D. That the employer develop a system to prepare, issue, and cancel entry permits

    Answer: D. That the employer develop a system to prepare, issue, and cancel entry permits

35. Which of the following is *not* a duty of the entrant:
    A. Properly use all assigned equipment
    B. Communicate with the attendant
    C. Exit when told to
    D. Continually test the level of toxic chemicals in the space

    Answer: D. Continually test the level of toxic chemicals in the space

36. Of the following, which is *not* a duty of the entry supervisor:
    A. Summon rescue services
    B. Terminate entry
    C. Remove unauthorized people
    D. Endorse the entry permit

    Answer: A. Summon rescue services

37. When designing ventilation systems for permit space entry,
    A. The air should be blowing into the space
    B. The air should always be exhausting out of the space
    C. The configuration, contents, and task determine the type of ventilation methods used
    D. Larger ducts and bigger blowers are better

    Answer: C. The configuration, contents, and tasks determine the type of ventilation methods used

38. Of the following, which is *not* a duty of the attendant:
    A. Know accurately how many entrants are in the space
    B. Communicate with entrants
    C. Continually test the level of toxic chemicals in the space
    D. Summon rescue services when necessary

    Answer: C. Continually test the level of toxic chemicals in the space

39. Circle the following true statement(s).
    A. Carbon monoxide gas should be ventilated from the bottom
    B. The mass of air going into a space equals the amount leaving
    C. Methane gas should be ventilated from the bottom
    D. Gases flow by the inverse law of proportion

    Answer: B. The mass of air going into a space equals the amount leaving

40. Hot work is going to be performed in a solvent reactor vessel that is 10 feet high and 6 feet in diameter. Which of the following is the preferred way to do this?
    A. Use submerged arc welding equipment
    B. Inert the vessel with nitrogen and provide a combination airline with auxiliary SCBA respirator for the welder
    C. Fill the tank with water and use underwater welding procedures
    D. Pump all the solvent out, ventilate for 24 hours, and use non-sparkling welding sticks
    E. Clean the reactor vessel, then weld per 1910.252

    Answer: E. Clean the reactor vessel, then weld per 1910.252

41. The certification of training required for attendants, entrants, and entry supervisors must contain (circle all that apply)
    A. The title of each person trained
    B. The signature or initials of each person trained

C. The signature or initials of the trainer
D. The topics covered by the training

Answer: C. The signature or initials of the trainer

42. Paragraph 1910.146 (g) requires that training of all employees whose work is regulated by the permit-required confined space standard shall be provided
    A. On an annual basis
    B. When the employer believes that there are inadequacies in the employee's knowledge of the company's confined space procedures
    C. When the union demands it
    D. All of the above.

    Answer: B. When the employer believes that there are inadequacies in the employee's knowledge of the company's confined space procedures

6.3 CONFINED SPACE TRAINING: THE BOTTOM LINE

Training is mandatory when an employee first is assigned confined space entry duties, when those duties change, whenever a change in permit-required confined space entry operations presents a new hazard, or whenever an employer believes an employee needs additional procedural assistance. The preceding point cannot be overstated. Here is another crucial point: The primary message sent by the employer in training his or her workers for confined space entry should be "look before you leap."

6.4 REFERENCE

The Office of the Federal Register. *Code of Federal Regulations Title 29 Parts 1900–1910 (.146)*, Office of Federal Register, Washington, D.C., 1995.

CHAPTER 7

Assignment of On-Site Personnel

Gradually the tunnel narrowed and darkened, and they needed their torches. The weight of the water around their ankles taught them a sliding gait, shuffling rather than raising their feet. Caught in the torch beams, rats scuttled along a jutting course of bricks, staring undaunted at the light, giving back loathing for loathing. When the little party stopped a chorus of echoing drips and trickling outfalls resounded in the cavernous space.

They reached the first weir after a few minutes. A bare iron ladder allowed them to climb over it single file, and stand in the first of Bazalgate's greater intercepting sewers. Unlike the sluggish water of the sleeping Fleet, this was rapidly flowing. Their torches showed an opaque brown fluid swirling past the ledge on which they stood. There was a very pervasive unhealthy smell.

They used to be able to light the streets of London with the methane coming off this . . . in the days of gas light, of course. Now if any of you gentlemen has a box of matches, or a cigarette lighter in your pockets, I must ask you not to touch them. The slightest spark can cause an explosion here. (Sayers & Walsh, Thrones, Dominations, *pp. 263–264, 1998)*

7.1 INTRODUCTION

ON-SITE personnel including entrants, attendants, and entry supervisors assigned to effect permit-required confined entries must be fully aware of their duties under the OSHA Standard, OSHA, under part (h) entrants, (i) attendants, and (j) entry supervisors of 1910.146, clearly defines these duties.

As stated in the previous chapter, training is the key ingredient to effecting safe permit-required confined space entry. Obviously, assigning anyone specific duties is easy, but ensuring that these duties are performed in the correct manner—especially when training has not been conducted—is much more difficult. Supervisors and workers must know their duties and must know how to complete their duties in a safe and correct manner.

7.2 DUTIES OF AUTHORIZED ENTRANTS

The key responsibility of any permit-required confined space entrant is to gain knowledge—knowledge of the hazards that may be faced during entry. The entrant must also be knowledgeable enough to understand the mode, signs or symptoms, and consequences of exposure to hazards (whatever they might be)—immediate or potential hazards. This knowledge requirement is central to the critical importance that training plays in compliance with this program—or any safety program.

Knowledge of the hazards and/or potential hazards is just part of the requirements involved in being a "qualified" entrant. The entrant must also know his or her equipment—how to use it, what it is to be used for, and its limitations.

The entrant must know how to communicate with the attendant. Communication can be via radio/walkie-talkie (which must also be intrinsically safe—capable of producing no sparks), hand signals (obviously visual contact must be maintained), and/or voice, whistle, or some other prearranged and practiced sound-making device.

The entrant must alert the attendant whenever he or she recognizes any warning sign or symptom of exposure to a dangerous situation. He or she must also communicate to the attendant any changing condition that could make the entry more hazardous than it already is.

The entrant must know when to exit the confined space—without hesitation, without prompting, without delay. He or she must maintain a position within the space whereby he or she can exit quickly if necessary. When ordered to exit by the attendant, the entrant must not delay, think about it, or pause for any reason—when ordered to exit, the entrant must exit *immediately.*

7.3 DUTIES OF THE ATTENDANT

The employer has the responsibility of ensuring that the permit-required confined space attendant is fully trained and knowledgeable. The attendant must know the hazards that may be faced during entry, including information on the mode, signs or symptoms, and consequences of the exposure. The attendant must be aware of the behavioral effects of hazard exposure to which the entrants may be subjected.

That the attendant plays a critical role in confined space entry should be apparent. This critical role cannot be filled by just anyone—the attendant must be fully trained and qualified to perform his or her assigned responsibilities.

The attendant is responsible for maintaining an accurate count of authorized entrants in the permit space and must ensure that the means used to identify authorized entrants accurately identifies who is in the permit space.

The attendant remains outside the permit space—until properly relieved by another qualified attendant. When the employer's permit entry program allows attendant entry for rescue, attendants may enter a permit space to attempt to rescue *if* they have been trained and equipped for rescue operations.

Note: Experience has shown that attendant entry rescue is not a good practice. If you look at the historical records dealing with fatalities in confined spaces, for every victim, there are usually additional fatalities—because of rescuers who become victims. Many people call this the John Wayne Syndrome; that is, rescuers or heroes rush into a confined space (like John Wayne) to rescue a victim—but unlike John Wayne, they do not end the rescue on a happy note, but instead become victims themselves. OSHA allows employers to train their personnel to make internal rescues—however, I recommend "external" rescue only.

The attendant maintains communication between himself/herself and the entrant(s) at all times. The attendant also monitors conditions within and outside the space that might endanger the entrants, and orders the entrants to exit if necessary.

If the attendant detects a hazardous situation, the behavioral effects of hazard exposure in an authorized entrant, and/or determines that he or she cannot (for whatever reason) perform his or her attendant duties, the attendant must order the immediate evacuation of the permit space.

The attendant is also responsible for summoning rescue and other emergency services as soon as he or she determines that authorized entrants may need assistance to escape from permit space hazards.

The attendant prohibits unauthorized people from entering a permit space and/or from interfering with an entry in progress.

The attendant has one responsibility and one responsibility only: To perform the duties of a permit space attendant without allowing distraction—any distraction.

7.4 DUTIES OF ENTRY SUPERVISORS

As with any other work activity, supervisors play a key role in permit-required confined space entry. In permit-required space entry, the supervisor is responsible for issuing confined space permits. To do this according to the standard, the entry supervisor must know the hazards of the confined spaces, verify that all tests have been conducted and all procedures and equipment are in place before endorsing a permit, terminate entry if necessary, cancel permits, and verify that rescue services are available and the means for summoning them are operable. In addition, entry supervisors are to remove unauthorized individuals who attempt to enter the confined space. They also must determine, at least when shifts and entry supervisors change, that acceptable conditions, as specified by the permit, continue.

Remember, the entry supervisor signs "the bottom line" on the permit. Before signing that bottom line on any safety document, supervisors should use good judgment, along with care and caution. If and when anything goes wrong in a confined space entry, the first item that the OSHA investigator will want to see is the permit. When lawyers are involved (as is often the case when workers are killed or badly injured on the job), the permit becomes an important document that will end up in a court of law—along with the supervisor in charge of the confined space entry operation.

7.5 REFERENCES

Sayers, D. L. & Walsh, J. L. *Thrones, Dominations.* New York: St. Martins Press, pp. 263–264, 1998.

The Office of the Federal Register. *Code of Federal Regulation Title 29 Parts 1900–1910 (.146),* Office of the Federal Register, Washington, D.C., 1995.

CHAPTER 8

Confined Space Rescue

Of the more than 1.6 million workers who enter confined spaces each year, approximately 63 die from asphyxiation, burns, electrocution, drowning and other tragedies related to confined space entry operations. But more alarming is the fact that 60 percent of those who die in confined spaces are untrained rescuers who not only fail to save a co-worker, but are killed during the rescue attempt (The John Wayne Syndrome). OSHA requires that a trained, equipped rescue team be available whenever employees work in confined spaces. (Coastal Video, Confined Space Rescue, *p. 2, 1993)*

8.1 INTRODUCTION

THE headline read "3 Honored for Fatal Rescue Try." The article, written by Jeff Long and appearing in the March 7, 1998, *Daily Press,* p. C10, went on to describe the incident that earned the three dead heroes their honors.

Nothing is new or unusual about this particular newspaper account. These accounts always seems to read the same way—a victim in a confined space, a topside hero (or heroes) who abandons common sense and concern for his or her own safety and enters a dangerous environment to save his or her co-worker and friend. Like the victim whose peril sparked the action, he or she also dies.

Those heroes chose to risk—and some of them lose—their lives when they enter a confined space that contains an immediate and apparent risk. The victims' deaths affect their families, their friends, and their co-workers.

Supervisors responsible for the safety of workers in dangerous jobs may face certain tasks that only become harder with repetition. Have you ever had a worker in your charge die on the job? Have you ever had to go to the victim's home and knock on the door? When you knock, that worker's happy, beautiful children race to open the door and are delighted to see you, because you wear the same uniform as their father or mother. They call back into the house that "somebody from work is here!"

Then to your worker's spouse (who must know, just by your presence, that something terrible has happened), you deliver a message—a message that will devastate that family, a message that will change their lives forever.

Have you been there? If you have, you know how much you would give to never have the need to make such a visit to a worker's home again.

8.2 RESCUE SERVICES

The employer who engages in permit-required confined space entry has the option of whether to use an off-site or in-plant rescue service. If the decision is made to use an off-site rescue service, a number of factors must be considered.

The first factor to consider is: Is such a rescue service readily available? This is a logical, straightforward question. However, when you seek to answer it, you may find that the question is easier to ask than it is to obtain an answer that ends your search. Why? Let's take a look at what typically occurs when this option is chosen.

The natural inclination is to list 911 or another local emergency number on your confined space permit as your rescue service. However, is such rescue service really available to you from the local fire department or some other emergency service? You need to find out. Typically, when we call our local fire departments and explain to them that we are about to make a confined space entry—that we are giving them a heads up to be aware of the operation—they are usually puzzled. "We fight fires and make some rescues. But, confined space rescue? Sorry, we are not trained for that." Have your heard this response before? If you've tried to locate off-site rescue service, you probably have, because that response is typical.

We simply cannot list 911 as the standby emergency rescue service (and hope that whoever responds will be able to effect rescue) unless we are absolutely certain they will respond in less than 4 minutes (remember, a victim in a confined space cannot live without air for more than 4 minutes) and are fully trained to effect the rescue.

The second factor that you must take into consideration (once you have identified a rescue service that can respond in 4 minutes or less) is: Is this service familiar with your facility? Have you invited the members of the service into your facility for familiarization?

Another factor to consider is on-site training. Has the rescue service actually practiced making confined space rescues in your confined spaces? Are they willing to spend the time to acquire the information they need to handle a crisis situation at your facility? This is an important point—one that an OSHA auditor will be certain to verify if and when your facility is audited.

On-site rescue teams have considerations as well. If you decide to employ the services of an on-site rescue team, OSHA requires the following:

(1) The employer shall ensure that each member of the rescue team is provided with, and is trained to use properly, the personal protective equipment and rescue equipment necessary for making rescues from permit spaces.

(2) Each member of the rescue team shall be trained to perform the assigned rescue duties. Each member of the rescue team shall also receive the training required of authorized entrants stated in the standard and described in section 7.2 of this text.

(3) Each member of the rescue team shall practice making permit space rescues at least once every 12 months, by means of simulated rescue operations in which they remove dummies, mannequins, or actual people from the actual permit spaces, or from representative permit spaces. Representative permit spaces shall, with respect to opening size, configuration, and accessibility, simulate the types of permit spaces from which rescue is to be performed.

(4) Each member of the rescue team shall be trained in basic first aid and in cardiopulmonary resuscitation (CPR). At least one member of the rescue team holding current certification in first aid and in CPR shall be available.

In the OSHA standard, the above requirements describe the rescue team as a "rescue service." From experience, calling this rescue service a rescue team is better (and more appropriate), because a team is what it is. To properly effect confined space rescue, the rescue service must be a "team"—individuals who work together seamlessly. Each member must have good endurance, enthusiasm, willingness to learn, and possess a team-oriented attitude.

8.2.1 RESCUE SERVICE PROVIDED BY OUTSIDE CONTRACTORS

When an employer arranges to have people other than the host employer's employees perform permit space rescue, the host employer shall

(1) Inform the rescue service of the hazards they may confront when called on to perform rescue at the host employer's facility

(2) Provide the rescue service with access to all permit spaces from which rescue may be necessary so that the rescue service can develop appropriate rescue plans and practice rescue operations

(a)

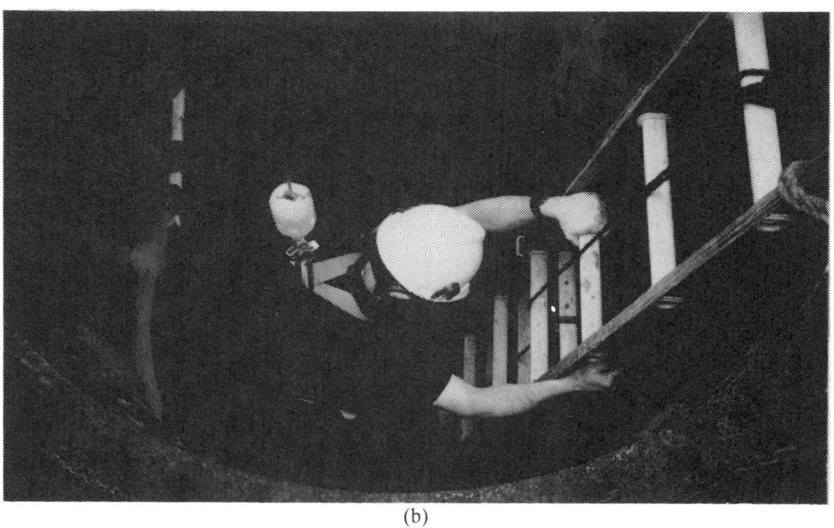

(b)

Figure 8.1 (a) Tripod, winch, and tending line. (b) Confined space entrant attached to a tripod mounted winch line, entering a permit-required confined space.

8.2.2 NON-ENTRY RESCUE

The rescue services we've discussed to this point all involved making external (non-entry rescue) confined space rescues—the preferred method of rescue recommended by this text, even though it may not be feasible on all occasions. The rule of thumb that I use is that if external rescue via a tripod, winch, retrieval line, and body harness (shown in Figure 8.1) cannot be made, then the confined space entry should not be made in the first place. When such retrieval systems are used, they shall meet the following requirements.

(1) Each authorized entrant shall use a chest or full body harness, with a retrieval line attached at the center of the entrant's back near shoulder level or above the entrant's head. Wristlets may be used in lieu of the chest or full body harness if the employer can demonstrate that the use of a chest or full body harness is unfeasible or creates a greater hazard and that the use of wristlets is the safest and most effective alternative.
(2) The other end of the retrieval line shall be attached to a mechanical device or fixed point outside the permit space in such a manner that rescue can begin as soon as the rescuer becomes aware that rescue is necessary. A mechanical device (such as a tripod and winch assembly) to retrieve personnel from vertical-type permit spaces more than 5 feet (1.52 m) deep.

A final word on permit-required confined space rescue: In the event of a rescue where the entrant is exposed to a hazardous material for which a Material Safety Data Sheet (MSDS) or other similar written information is required to be kept at the work site, that MSDS or written information must be made available to the medical facility treating the exposed entrant.

8.3 REFERENCES

Coastal Video. *Confined Space Rescue Booklet.* Virginia Beach, VA: Coastal Video Communication Corp., 1993.

The Office of the Federal Register. *Code of Federal Regulations Title 29 Parts 1900–1910 (.146),* Office of the Federal Register, Washington, D.C., 1995.

CHAPTER 9

Alternative Protection Measures

Minimizing the amount of regulation that applies to spaces whose hazards have been eliminated encourages employers to actually remove all hazards.

9.1 INTRODUCTION

OSHA has specified alternative protection procedures that may be used for permit spaces where the only hazard is atmospheric and ventilation alone can control the hazard. Let's take a brief look at these alternative protection procedures.

9.2 "HIERARCHY" OF PERMIT-REQUIRED CONFINED SPACE ENTRY

The following hierarchy of permit-required confined space entry is useful to anyone involved in designing a work site confined space entry program.

(1) 1910.146 (c) (7) Reclassification—Hazards *Eliminated*
 *Requires certification, (c) (7) (iii)
 A permit-required confined space may be reclassified by the employer as a non-permit confined space under the following procedures:
 - If the permit space poses no actual or potential atmospheric hazards and if all hazards within the space are eliminated without entry into the space, the permit space may be reclassified as a non-permit confined space for as long as the non-atmospheric hazards remain eliminated.
 - If it is necessary to enter the permit space to eliminate hazards, such entry shall be performed under the guidelines presented in the standard. If testing and inspection during that entry demonstrate that the hazards within the permit space have been elimi-

nated, the permit space may be reclassified as a non-permit confined space for as long as the hazards remain eliminated.

Note: OSHA points out that control of atmospheric hazards through forced air ventilation does not constitute elimination of the hazards.

- The employer shall document the basis for determining that all hazards in a permit space have been eliminated, through a certification that contains the date, the location of the space, and the signature of the person making the determination. The certification must be made available to each employee entering the space.

Note: Great care and caution should be exercised before anyone signs certification stating that a particular confined space is not hazardous (for any reason). Remember, the person who signs such a document is responsible, and therefore liable, for his or her decision.

- If hazards arise within a permit space that has been declassified to a non-permit space, each employee in the space must exit the space. The employee must then reevaluate the space and determine whether it must be reclassified as a permit space.

(2) 1910.146 (c) (5) (i) (E) Alternative Entry—Hazards *Controlled* (by continuous forced air ventilation)

- requires documentation and supporting data, (c) (5) (i) (E)
- requires training, (g)
- requires a "mini-program," (c) (5) (ii)
- requires certificate, (c) (5) (ii) (H)

An employer may use the alternate procedures specified in the standard (c) (5) (ii) for entering a permit space under the conditions set forth in the following.

- The employer can demonstrate that the only hazard posed by the permit space is an actual or potential hazardous atmosphere.
- The employer can demonstrate that continuous forced air ventilation alone is sufficient to maintain that permit space safe for entry.
- The employer develops monitoring and inspection data that support his or her reclassification decision.
- If an initial entry of the permit space is necessary to obtain the data required, the entry must be made by the requirements set forth for entry into a permit-required confined space.
- The determinations and supporting data are documented by the employer and are made available to each employee who enters the permit space.

Let's summarize these requirements. To qualify for alternative procedures, employers must

- ensure that it is safe to remove the entrance cover (e.g., a manhole filled with methane might explode if, when removing the metal manhole cover, the cover and/or tools used cause a spark)

- determine that ventilation alone is sufficient to maintain the permit space safe for entry and that work to be performed within the permit-required space will introduce no additional hazards
- gather monitoring and inspection data to support the above requirements
- if entry is necessary to conduct initial data gathering, perform such entry under the full permit program
- document the determination and supporting data, and make them available to employees

(3) Permit Space Entry—Hazards cannot be eliminated nor controlled. If this is the case, then the following are required:
- written program, (d) as required by (c) (4)
- permits, (e) and (f)
- training, (g)
- attendant, (d) (6)
- testing, (d) (5)
- rescue, (k)

Maintaining your confined space entry program (once it is established for your facility) will require regular attention on your part in evaluation and analysis of your facilities and their spaces. As your facility changes, grows, and ages, the confined spaces on your site may change, too, demanding reassessment. Meeting the requirements of OSHA's standards for your facility is an ongoing process, not a once-and-done event.

9.3 REFERENCE

The Office of the Federal Register. *Code of Federal Regulation Title 29 Parts 1900–1910 (.146)*, Office of the Federal Register, Washington, D.C., 1995.

CHAPTER 10

Procedures for Atmospheric Testing

You can never trust your senses to determine if the air in a confined space is safe! You cannot see or smell many toxic gases and vapors, nor can you determine the level of oxygen present.

10.1 INTRODUCTION

PERSONNEL involved in permit-required confined space entry must understand that some vapors or gases are heavier than air and will settle to the bottom of a confined space. Other gases are lighter than air and will be found around the top of the confined space. Because of the behaviors of various toxic gases, you must test all areas (top, middle, bottom) of a confined space with properly calibrated testing instruments to determine what gases are present (see Figure 10.1).

In this chapter, we cover OSHA's requirements and procedures for atmospheric testing.

10.2 TESTING PROCEDURES

Atmospheric testing is required for two distinct purposes: evaluation of the hazards of the permit space and verification that acceptable entry conditions for entry into that space exist.

(1) *Evaluation testing:* The atmosphere of a confined space should be analyzed using equipment of sufficient sensitivity and specificity to identify and evaluate any hazardous atmospheres that may exist or arise, so that appropriate permit entry procedures can be developed and acceptable entry conditions stipulated for that space. Evaluation and interpretation of these data, and development of the entry procedure, should be done by, or reviewed by, a technically qualified professional

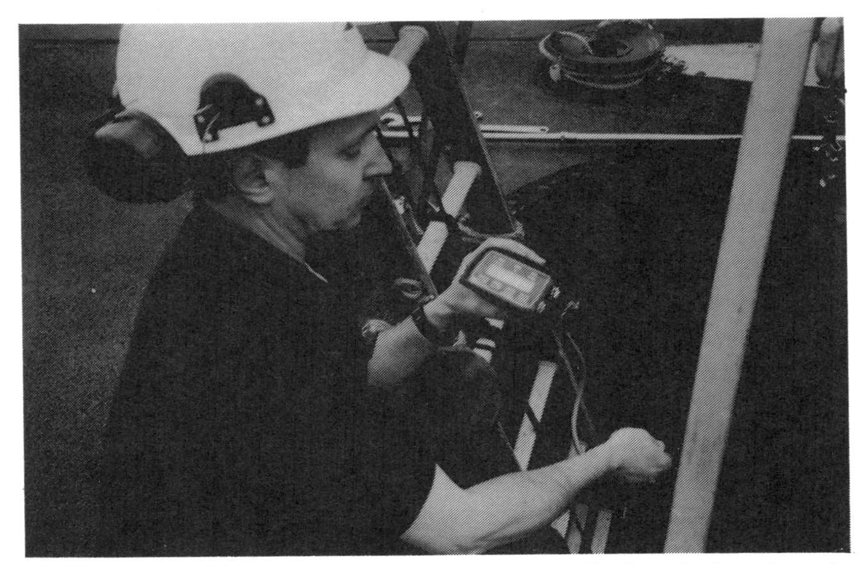

Figure 10.1 Entry supervisor checking readings on air monitor. The air monitor's sensing probe is being lowered a step at a time to measure top, middle, and lowest levels for contaminants.

(e.g., OSHA consultation service, Certified Safety Professional (CSP), Certified Industrial Hygienist (CIH), registered safety engineer, etc.) based on evaluation of all serious hazards.

(2) *Verification testing:* The atmosphere of a permit space that may contain a hazardous substance should be tested for residues of all contaminants identified by evaluation testing using permit-specified equipment to determine that residual concentrations at the time of testing and entry are within the range of acceptable entry conditions. Results of testing (i.e., actual concentration, etc.) should be recorded on the permit in the space provided adjacent to the stipulated acceptable entry condition.

(3) *Duration of testing:* Measurement of values for each atmospheric parameter should be made for at least the minimum response time of the test instrument specified by the manufacturer.

(4) *Testing stratified atmospheres:* When monitoring for entries involving a descent into atmosphere that may be stratified, the atmospheric envelope should be tested a distance of approximately 4 feet (1.22 m) in the direction of travel and to each side. If a sampling probe is used, the entrant's rate of progress should be slowed to accommodate the sampling speed and detector response.

10.3 AIR MONITORING AND OSHA

When an OSHA compliance officer audits your facility, if you have permit-required confined spaces that are entered by your employees, the auditor will pay particular attention to your air monitoring procedures.

Typically, the OSHA auditor will want to see copies of your confined space permits for the past year. From these permits, the auditor will choose one and set it aside. Later, the auditor will ask to interview those involved in making that confined space entry. The auditor may ask the confined space personnel several different questions related to their knowledge of confined space entry. The auditor may desire to see these personnel perform the entry again (if possible).

During the OSHA auditor's interview process, air monitoring will be discussed. The auditor will want to see the instrument used during the confined space entry. The auditor will note the condition of the instrument, looking specifically for any damage, dirt, or dead batteries (Are they using the right batteries or have they "jury-rigged" a battery pack?), and will test to determine if any sensors are malfunctioning, etc.

The OSHA auditor always asks one of the confined space entry personnel to demonstrate both how to calibrate and how to use the instrument.

In addition, the OSHA auditor typically asks several questions related to air monitoring, to determine the knowledge level of the workers. The following questions are normally asked:

(1) Have the operators been trained?
(2) Who gave them training? What was covered? How long did the training last? Any hands-on or on-the-job training?
(3) What type of instruments are used?
(4) Where is the manufacturer's instruction manual? Have they read the manual?
(5) How often do they use the instrument?
(6) Do they have calibration data, logbook, etc.?
(7) What calibration gas do they use? Why did they choose this gas (a question mainly for percent LEL—are they using propane, methane, pentane, etc.)?
(8) Do they zero the instrument as part of the calibration?
(9) Who calibrates the equipment? How often? How is it done?
(10) Do they have a calibration curve or correction factor chart?
(11) What are the interferences for the toxic sensors?
(12) Is the meter intrinsically safe for the environment they are monitoring?

(13) Are they waiting long enough for the sensors to respond (and for remote sampling—some manufacturers suggest 1 second per foot of tubing)?
(14) Are they testing all levels and areas where entrants will be working?
(15) If using several individual instruments, are they testing in the right sequence (oxygen, flammables, toxics)?
(16) What do the numbers on the instrument mean? Are they exact?
(17) What are you comparing the numbers to? What is considered safe for entry?
(18) Have they replaced any sensors? Any batteries? Any other parts? Do they have maintenance logs?
(19) Do they send the instrument back to manufacturer on a regular basis for complete calibration and maintenance?
(20) Do they field check?

Note: If you use your portable gas detector for sewer entry, the OSHA auditor will check your detector to see if it complies with OSHA's May 19, 1994, technical amendments to the confined space rule CFR 1910.146, where Federal OSHA modified the Appendix E language to read as follows:

> The oxygen sensor/broad range sensor is best suited for initial use in situations where the actual or potential contaminants have not been identified, because *broad range* sensors, unlike *substance-specific* sensors, enable employers to obtain an overall reading of the hydrocarbons (flammables) present in the space.

10.4 OTHER OSHA PERMIT-REQUIRED CONFINED SPACE AUDIT ITEMS

In section 10.3, we discussed the types of queries that an OSHA auditor would make concerning a typical work site's air monitoring practices used in performing permit-required confined space entry. In this section, we discuss the "other" OSHA audit items—ones dealing specifically with permit-required confined space entry procedures.

When an OSHA auditor audits your confined space entry program, you can be assured that he or she will look at most (if not all) of the items listed below.

(1) Are aisles in the vicinity of the confined space marked?
(2) Are aisles and passageways properly illuminated?
(3) Are aisles kept clean and free of obstructions?
(4) Are fire aisles, access to stairways, and fire equipment kept clear?
(5) Is there safe clearance for equipment through aisles and doorways?

(6) Have all confined spaces and permit-required confined spaces been identified?
(7) Are danger signs posted (or other equally effective means of communication) to inform employees about the existence, location, and dangers of permit-required confined spaces?
(8) Is the written permit-required confined space entry program available to employees?
(9) Is the permit-required confined space sufficiently isolated? Have pedestrian, vehicle, or other necessary barriers been provided to protect entrants from external hazards?
(10) When working in permit-required confined spaces, are environmental monitoring tests taken and means provided for quick removal of workers in case of an emergency?
(11) Are confined spaces thoroughly emptied of any corrosive or hazardous substances (such as acids or caustics) before entry?
(12) Are all lines to a confined space containing inert, toxic, flammable, or corrosive materials valved off and blanked or disconnected and separated before entry?
(13) Is each confined space checked for decaying vegetation or animal matter that may produce methane?
(14) Is the confined space checked for possible industrial waste that could contain toxic properties?
(15) Before permit space entry operations begin, has the entry supervisor, who is identified on the permit, signed the entry permit to authorize entry?
(16) Has the permit been made available at the time of entry to all authorized entrants (by being posted at the entry portal or by other equally effective means) so that entrants can confirm that pre-entry preparations have been completed?
(17) Is necessary personal protective equipment (PPE) available?
(18) Has necessary lighting equipment been provided?
(19) Has equipment (such as ladders) needed for safe ingress and egress by authorized entrants been provided?
(20) Is rescue and emergency services equipment available?
(21) Is it required that all agitators, impellers, or other rotating equipment inside confined spaces be locked out if they present a hazard?
(22) Is all portable electrical equipment used inside confined spaces either grounded and insulated, or equipped with ground fault protection?
(23) Is at least one attendant stationed outside the confined space for the duration of the entry operation?

(24) Is there at least one attendant whose sole responsibility is to watch the work in progress, sound an alarm if necessary, and render assistance?

(25) Is the attendant trained and equipped to handle an emergency?

(26) Is the attendant and/or are other workers prohibited from entering the confined space without lifelines and respiratory equipment if there is any question as to the cause of an emergency?

(27) Is communications equipment provided to allow the attendant to communicate the authorized entrants as necessary to monitor entrant status and to alert entrants of the need to evaluate the permit space?

If your worker training brings your workers to the level that they can provide reasonable answers for the sample OSHA questions, and if your facility is compliant with the list of OSHA auditor questions, you are well on your way to a successful confined space entry program. Chapter 11 provides an effective sample of a written confined space entry program.

10.5 REFERENCE

The Office of the Federal Register. *Code of Federal Regulations Title 29 Parts 1900–1910 (.146)*, Office of the Federal Register, Washington, D.C., 1995.

CHAPTER 11

Written Permit-Required Confined Space Entry Program

In any instance of a confined space entry, the intent of management is that most appropriate and safe confined space working procedures are to be followed. In this intent, management's support by the availability of personal safety equipment and a standardized procedure to ensure the safety of its employees shall not be compromised by any responsible or assigned supervisory representative. (WEF, p. 5, 1994)

11.1 INTRODUCTION

WITHOUT strong management support and backing, a work site's safety program cannot function as it is intended and as required by law. Management's support of work site safety programs includes many elements. Management must be ultimately responsible—and accountable—for worker and work site safety. Management must provide the necessary financing to support the work site safety program. Management must provide for proper employee training. Management must support and enforce compliance with standards, rules, and regulations applicable to a safe work environment.

Because we are addressing a topic that directly impacts all work site employees' well-being, health, and safety, the importance of management's commitment to supporting the work site safety program cannot be overstated.

For permit-required confined space entry, management must make a commitment via managerial policy. This policy must include written procedures and directions, and should be preceded by a strong policy statement such as the one provided by the Water Environment Federation, which opens this chapter.

OSHA, in Appendix C of 1910 Standard, provides sample written confined space programs. The examples provided by OSHA in the standard are for work in sewers, meat and poultry rendering plants, and for work places where tank cars, trucks and trailers, dry bulk tanks and trailers, railroad tank cars, and similar portable tanks are fabricated or serviced.

This chapter provides a confined space pre-entry checklist and a written confined space entry program. The checklist and written programs are also samples, adaptable to your own facility's needs. Note that the example confined space program presented in this chapter has been used in wastewater/water treatment and other general industry locations for approximately 20 years—well before OSHA had a requirement for a formal confined space standard.

While no single written program is the "best" program, the program illustrated here has a huge advantage over many others: It has been tested. It has proven its worth by surviving direct OSHA scrutiny. The sample confined space program presented here has been amended many times and has been constantly updated to reflect changes that have been made in the OSHA standard.

11.2 CONFINED SPACE PRE-ENTRY CHECKLIST

Before initiating a permit-required confined space entry, certain pre-entry checks should be made first. In the following, many suggested pre-entry checks are listed. Each work site should develop its own site-specific pre-entry checklist.

11.2.1 PRE-ENTRY CHECKLIST FOR PERMIT-REQUIRED CONFINED SPACE ENTRY (A SAMPLE)

Do *not* enter a confined space until you have considered every question and have determined the space to be safe to be entered.

(1) Is entry necessary?
(2) Are the instruments used in atmospheric testing properly calibrated?
(3) Was the atmosphere in the confined space tested?
(4) Was oxygen at least 19.5 percent—and not more than 23.5 percent?
(5) Were toxic, flammable, or oxygen-displacement gases/vapors present?
(6) Will the atmosphere in the space be monitored while work is going on? Continuously? Periodically?
(7) Has the space been cleaned before entry?
(8) Has the space been ventilated before entry?

(9) Will ventilation be continued during entry?
(10) Is the air intake for the ventilation system located in an area that is free of combustible dusts, vapors, and toxic substances?
(11) If atmosphere was found unacceptable and then ventilated, was it re-tested before entry?
(12) Has the space been isolated from other systems?
(13) Has electrical equipment been locked out?
(14) Has mechanical equipment been blocked, chocked, and disengaged where necessary?
(15) Have lines under pressure been blanked and bled?
(16) Is special clothing required?
(17) Is rescue equipment and/or communications equipment required?
(18) Are spark-proof tools required?
(19) Is respiratory protection required?
(20) Have you been trained in proper use of a respirator?
(21) Have you received first aid/CPR training?
(22) Have you been trained in confined space entry, and do you know what dangers to look for?
(23) Will there be a standby person on the outside in constant visual or auditory communication with the person on the inside?
(24) Has the standby person been trained in rescue procedures?
(25) Are you familiar with emergency rescue procedures?
(26) Has a confined space entry permit been issued?
(27) Does the permit include a list of emergency telephone numbers?

11.3 WRITTEN CONFINED SPACE PROGRAM (A SAMPLE)

I. PURPOSE

This confined space entry program has been developed in accordance with the Occupational Safety and Health Administration (OSHA) regulations 29 CFR 1910.146. The purpose of this program is to ensure that proper protection is taken for all employees working in confined spaces.

II. GENERAL PROGRAM MANAGEMENT

A. Responsibility

It is the responsibility of management to protect the organization's employees. Specifically, organizational management will

(1) Evaluate the work place to determine which spaces are permit-required confined spaces
(2) Inform work place employees of the permit-required confined spaces
(3) Determine whether employees will enter permit spaces and what effective measures will be taken to prevent employees from entering permit spaces
(4) Require employees who will enter permit spaces to use a fully completed confined space permit

The organization's safety director is responsible for this program and has authority to make decisions to ensure the success of this program. Copies of the written program are installed in Safety Standard Operating Procedures (SOPs), and additional copies may be obtained from the safety office.

B. Program Review and Update

The confined space program will be reviewed and/or updated under the following circumstances:

(1) When the Safety Division has reason to believe that measures taken under the confined space program may not protect employees. In such cases, the program will be revised to correct deficiencies before authorizing subsequent entries, and the permit-required confined space program will be reviewed, using canceled permits retained within 1 year after each entry, and the revised program will be as necessary.

C. Identification of Confined Spaces

The confined space entry program requires identification of all confined spaces within the organization.

(1) Each department shall, with assistance from Safety Division personnel, be responsible for identification of all potential confined spaces in their facilities that employees are subject to entering. The confined space list will be prominently displayed on work center bulletin boards. Any new or existing process, vessel, or space meeting the cri-

teria based on engulfment and hazardous atmosphere of the confined space entry program shall be identified.

(2) Confined spaces within the organization include, but are not limited to, manholes, vertical entry spaces, channels, covered tanks, grit tanks, clarifiers, aeration tanks, contact tanks, thickeners, digestors, incinerators, chemical tanks, scrubber towers, pumping station wet wells, and water meter vaults.

(3) After confined space designation, a confined space entry permit shall be filled out specific to each space and shall be completed only by the qualified person designated by each department.

(4) Warning signs shall be placed in well-traveled areas to promote awareness and to prevent unauthorized entry into potential confined spaces. The warning signs shall be issued by the work center supervisors and shall contain the warning shown in Figure (3.2) or similar warning.

Note: Warning signs are *not* required for manholes; however, *all* vertical entry manholes *are* confined spaces—entry by permit only!

III. METHODS OF COMPLIANCE

A. General Requirements

Confined space entry must be performed in accordance with the following requirements:

(1) Any condition making it unsafe to remove an entrance cover will be eliminated before the cover is removed.

(2) When covers are removed, the entrance will be promptly guarded by a barrier that will prevent accidental fall through the opening and will protect employees in the space from foreign objects entering the space.

(3) Before an employee enters the space, the internal atmosphere will be tested with a calibrated direct-reading instrument, for the following conditions in the order given:

a. Oxygen content

b. Flammable gases and vapors

c. Potential toxic air contaminants (see Appendix A)

In addition, the space will be thoroughly checked for the possibility for engulfment. All necessary steps will be taken to guard against engulfment, including lockout/tagout, double-valve protection, blinding and blanking, etc.

(4) There may be no hazardous atmosphere within the space whenever any employee is inside the space.

(5) Continuous forced air ventilation will be used as follows:
  a. An employee may not enter the space until forced air ventilation has eliminated any hazardous atmosphere.
  b. Forced air ventilation will be directed to ventilate the immediate areas where an employee is or will be, and will continue until all employees have left the space.
  c. The air supply for the ventilation will be clean and may not increase the hazard.

(6) The atmosphere within the space will be periodically tested as necessary to ensure that the continuous forced air ventilation is preventing the accumulation of a hazardous atmosphere.

(7) If a hazardous atmosphere is detected during entry
  a. Each employee will leave the space immediately.
  b. The space will be evaluated to determine how the hazardous atmosphere developed.
  c. Measures will be implemented to protect employees from the hazardous atmosphere before a subsequent entry.

(8) Before each entry, the affected work center will verify that the space is safe for entry and that the measures above have been taken, with a written certification giving the date, location of the space, and signature of the person providing the certification.

The organization may use alternate procedures for entering a permit-required space providing that

(1) It can be demonstrated that the only hazard is actual or potentially hazardous atmosphere
(2) It can be demonstrated that forced air ventilation alone is sufficient to maintain safe entry
(3) Monitoring and inspection data are developed
(4) If an initial entry is needed to collect the data above, then it will be performed in compliance with parts (C) to (I) of this program
(5) The determination and data required above are to be documented and available to employees who enter the space

Note: Employees entering a permit space using alternate procedures need not comply with the following parts of this program:

- permit-required confined space (C)
- permit system (D)
- entry permit (E)
- duties of attendants (G)
- duties of supervisors (H)
- rescue and emergency services (I)

B. Non-Permit Space

A space classified as a permit-required space may be reclassified as a non-permit space:

(1) If the permit space poses no actual or potential atmosphere hazards and if all hazards are eliminated without entering the space, it can be reclassified as a non-permit space as long as the non-atmospheric hazards remain eliminated.
(2) If it is necessary to enter the permit space to eliminate hazards, such entry will be performed under (C) and (I) of this program. If testing and inspection demonstrate that the hazards have been eliminated, the permit space can be reclassified as a non-permit space for as long as the hazards remain eliminated.
(3) The organization's Safety Division will document the basis for determining that all hazards have been eliminated through a certification that contains the date, location of the space, and the signature of the person making the determination.
(4) If hazards arise within a permit space that has been declassified to a non-permit space, all employees must exit the space. The space will be re-evaluated to determine if it must be reclassified as a permit space.

C. Permit-Required Confined Spaces

(1) The organization has implemented measures necessary to prevent unauthorized entry into a confined space.
(2) The organization will identify and evaluate the hazards of the permit spaces before employees enter them.
 Note: This requires atmospheric testing with a gas detector before entry into the space.
(3) Each affected organization work center will provide the following equipment to employees at no cost, maintain the equipment properly, and ensure that employees use that equipment properly:
 a. Testing and monitoring equipment needed to evaluate the permit space conditions
 b. Ventilation equipment needed to obtain acceptable entry conditions
 c. Communications equipment necessary for compliance
 d. Personal protective equipment insofar as feasible
 e. Lighting equipment needed to enable safe work in and exit from the space
 f. Barriers and shields to protect entrants from external hazards
 g. Equipment needed for safe ingress and egress

h. Rescue and emergency equipment to comply with this program
i. Any other equipment necessary for safe entry and rescue

(4) At least one attendant will be provided outside the permit space for the duration of entry operations.
(5) If multiple spaces are to be monitored by a single attendant, the means and procedures to enable the attendant to respond to an emergency in one or more spaces without distraction from the attendant's responsibility under this program shall be provided, which may require use of one attendant per entrant.
(6) The people who are to have active roles in entry operations are to be designated along with their duties; each member will be properly trained.
(7) Each work center will ensure that the on-site rescue team has been notified of confined space entry activity and is placed on standby alert for possible use; unauthorized personnel are to be prevented from attempting rescue.
(8) When more than one employer is involved in confined space entry, the work center supervisor will ensure that entry procedures are coordinated to ensure that they do not endanger each other.
(9) Whenever entry into a permit required space is required, the proper permit shall be used. These permits are to be properly prepared, issued, used, and canceled when entry is complete. Confined space entry permits must be retained on file by the work center supervisor for at least 1 year.

In addition to complying with requirements that apply to the organization, each contractor who performs permit space entry will

(1) Obtain any available information regarding permit space hazards and entry operations from the organization
(2) Coordinate entry operations with the organization when both organization and contractor personnel work in or near permit spaces, as required in this program
(3) Inform the organization of the permit space program that the contractor will follow, and any hazards confronted or created in permit spaces

D. Permit System

The confined space entry program requires issuance of a confined space permit prior to entry into a confined space. The purpose of the permit is to ensure safe working conditions prior to and during confined space entry.

Each work center will designate the "qualified person" responsible for the issuance and verification of the confined space entry permit and performance of all required atmospheric testing. The list of authorized "qualified people" for each department is attached as Appendix B to this program.

A confined space entry permit shall be issued prior to each confined space entry. A new permit shall be issued when

(1) Work within the confined space exceeds 12 continuous hours
(2) A power outage occurs
(3) Conditions change that could affect the safety of the entrant; i.e., the bleed line shows failure of the first block; oxygen, flammable, or toxic concentrations increase above acceptable limits, etc.

The confined space permit will be available for inspection at the work site while work is in progress.

After the permit has expired, it shall be filed in the work center for 1 year.

E. Confined Space Entry Permit

The entry permit used by the organization identifies

(1) The space to be entered
(2) The purpose of the entry
(3) The date and authorized duration of the entry
(4) The authorized entrants
(5) The personnel serving as attendants
(6) The individual serving as the entry supervisor (qualified person)
(7) The hazards of the permit space to be entered
(8) The measures used to isolate the space and eliminate or control hazards before entry
(9) The acceptable entry conditions
(10) The results of initial and periodic tests performed below (accompanied by the names or initials of the testers and by an indication of when the tests were performed):
   a. Test conditions in the permit space to determine if acceptable entry conditions exist before entry is authorized to begin, except that, if isolation of the space is not feasible because the space is larger or is part of a continuous system (such as a sewer), pre-entry testing will be performed to the extent feasible before entry is authorized, and if entry is authorized, entry conditions will be continuously monitored in the areas where authorized entrants are working.

b. Test or monitor the permit space as necessary to determine if acceptable entry conditions are being maintained during the course of entry operations.

c. When testing for atmospheric hazards, test first for oxygen, then for combustible gases and vapors, then for toxic gases and vapors.

Note: The test for oxygen must be done first, because most combustible gas sensors require oxygen to function.

(11) The on-site rescue team that can be called

(12) The communication procedures used by entrants and attendants to maintain contact with each other

(13) Equipment (such as testing equipment) to be provided for compliance with 29 CFR 1910.146

(14) Any other information necessary to ensure employee safety

(15) Any additional permits (such as hot work permits) issued for work in the space

F. Training

(1) The organization's Safety Division will provide training so that employees acquire understanding, knowledge, and skills necessary for the safe performance of the duties assigned.

(2) Training will be provided

a. Before the employee performs duties under this program

b. Before there is a change in permit space operations that presents a hazard about which the employee has not previously been trained

c. Whenever the qualified person has reason to believe there are inadequacies in their knowledge of these procedures

(3) The organization will certify that the training required has been accomplished and the employee is proficient in the duties; training records will be maintained by Safety Division/affected work center and Human Resources.

G. Duties of the Authorized Entrants

The organization shall ensure that all authorized entrants:

(1) Know the hazards that may be faced during entry

(2) Know how to use the equipment required

(3) Communication with the attendant as necessary to enable the attendant to monitor the entrants and to enable the attendant to alert the entrants of the need to evaluate as required

(4) Alert the attendant whenever
   a. The entrant recognized any warning sign or symptom of exposure to a dangerous situation
   b. The entrant detects a prohibited condition
(5) Exit from the permit space as quickly as possible whenever
   a. An order to evacuate is given by the attendant or supervisor
   b. The entrant recognizes any warning sign or symptom of exposure to a dangerous situation
   c. The entrant detects a prohibited condition
   d. An evacuation alarm is activated

H. Duties of Attendants

The organization will ensure that each attendant

(1) Knows the hazards that may be faced during entry
(2) Is aware of possible behavioral effects of hazard exposure
(3) Continuously maintains an accurate count of entrants
(4) Remains outside the permit space during entry until relieved by another attendant
(5) Communicate with entrants as necessary to monitor their status and to alert them of the need to evacuate
(6) Monitors activities inside and outside the space to determine if it is safe and orders evacuation immediately under any of the following conditions
   a. If the attendant detects any prohibited condition
   b. If the attendant detects the behavioral effects of hazard exposure in an entrant
   c. If the attendant detects a situation outside the space that could endanger entrants
   d. If the attendant cannot effectively and safely perform all duties required
(7) Summons on-site rescue and other emergency services (911) as soon as it is determined that entrants may need assistance to escape
(8) Prevents unauthorized people from approaching or entering a permit space while entry is underway
(9) Performs non-entry (external only) rescue using the safety hoist-winch assembly
(10) Performs no duties that might interfere with the attendant's primary duty to monitor and protect the authorized entrants

I. Duties of Qualified Person (Entry Supervisor)

Each qualified person shall:

(1) Know the hazards that may be faced during entry
(2) Verify (by checking that the appropriate entries have been made on the permit) that all tests specified by the permit have been conducted and that all procedures and equipment specified by the permit are in place before endorsing the permit and allowing entry to begin
(3) Terminate the entry and cancel the permit as required by this program
(4) Verify that on-site rescue services are available and that the means for summoning them are operable
(5) Remove unauthorized individuals who enter or attempt to enter the permit space during operations
(6) Determine that entry operations remain consistent with terms of the entry permit and that acceptable entry conditions are maintained

J. Rescue and Emergency Services

(1) The organization will provide trained confined space non-entry rescue personnel for each permit-required confined space entry affected.
   a. These rescue personnel must be on site, must be aware that confined space entry is being performed, and must be immediately available to perform rescue.
   b. Members of confined space rescue teams will be fully equipped with proper PPE. They will also receive the training required of authorized entrants under this plan (to ensure they have at least the level of knowledge required of entrants).
   c. Each member of the rescue team will practice making non-entry rescue.
   d. Each member of the rescue team will be trained in basic first aid and CPR.
(2) To facilitate non-entry (external) rescue, retrieval systems will be used whenever an authorized entrant enters a permit space, unless this would increase risk or would not assist the rescue. Retrieval systems will meet the following:
   a. Each authorized entrant will use a D-Ring full body harness with a static retrieval line.
   b. The other end of the retrieval line will be attached to a mechanical

device or fixed point outside the permit space so that rescue can begin as soon as it becomes necessary.

(3) If an injured entrant is exposed to a substance for which an MSDS or other similar written information is required, that sheet or written information will be made available to the medical facility treating the exposed entrant.

APPENDIX A

Chemical Exposure Limit List

Chemical Name	PEL[2] (8-hr. avg)	MEL[3]	IDLHL[4]
1. Ammonia	50 ppm	—	300 ppm
2. Carbon dioxide[5]	5,000 ppm	—	40,000 ppm
3. Carbon monoxide[5]	50 ppm	—	1,200 ppm
4. Sodium hydroxide	MUST USE SCBA TO ENTER		
5. Sulfur dioxide[5]	5 ppm	—	100 ppm
6. Chlorine[5]	1 ppm	1 ppm	10 ppm
7. Hydrogen chloride	5 ppm	5 ppm	50 ppm
8. Hydrogen sulfide	10 ppm	20 ppm	100 ppm
9. Propane[5]	1,000 ppm	—	2,100 ppm
10. Oxygen	19.5% (min)	23% (max)	—
11. Flammable	10% LEL	—	—

Note: The safety director may be contacted for limits on other substances as necessary.

APPENDIX B

Department Qualified Person List

Treatment Department
Plant Manager/Superintendent
Chief Operators

[2]PEL—Permissible exposure limit.
[3]MEL—Maximum exposure limit.
[4]IDLHL—Immediately dangerous to life or health limit.
[5]For determining safe entry into spaces containing these substances, Drager detection tubes are available. Contact your supervisor. If unusual order is apparent, do not enter space until proper identification is made.

Lead Operators
Designated, Fully Trained Operators

Engineering Department
Electrical Superintendent
Senior Instrument Specialist
Electrician Supervisor
Instrument Specialist
Electricians
Lead Mechanic
Carpenter/Painter
Diesel Mechanic

Interceptors
System Superintendents
Chief Foremen
Pump Station Supervisors
Systems Foremen
System Reliability Manager
Interceptor Supervisor
Reliability Manager
Portable Equipment Techs
Chemical Techs
Interceptor System Techs
System Service Techs
System Chemical Techs
System Maintenance Techs
Reliability Techs
Reliability Supervisor

Water Quality
Industrial Waste Managers
Industrial Waste Techs
Industrial Waste Specialists
Industrial Waste Assistants
Technical Services Techs/Specs
Environmental Scientists
Stormwater Specialists/Assistants

OSHA, as I have said repeatedly, requires exacting and detailed paperwork to support your safety programs and training. This sample confined space entry program can go a long way toward fulfilling your requirements for this particular OSHA standard. However, several other OSHA concerns are critically important to your confined space entry program as well—lockout/tagout, respiratory protection, and hot work permits, which are covered in the remaining chapters of this book.

11.4 REFERENCES

The Office of the Federal Register. *Code of Federal Regulations Title 29 Parts 1900–1910 (.146)*, Office of the Federal Register, Washington, D.C., 1995.

Water Environment Federation, *Confined Space Entry*. Alexandria, VA: Water Environment Federation, 1994.

11.5 RECOMMENDED REFERENCES FOR CONFINED SPACE ENTRY

The following publications are available from American Industrial Hygiene Association Publications, 2700 Prosperity Avenue, Suite 250, Fairfax, VA 22031, (703) 849-8888.

Confined Space Entry: An AIHA Protocol Guide, Vernon Rose & Terry King, ISBN 0-932627-67-6, 1995, 53 pages.

Manual of Recommended Practice for Combustible Gas Indicators and Portable Direct-Reading Hydrocarbon Detectors, 2nd ed., C. F. Chelton, ISBN 0-932627-48-X, 1993, 55 pages.

Direct-Reading Colorimetric Indicator Rubes Manual, 2nd ed., Janet Perper & Barbara Dawson, 1993, 60 pages.

The following book can be purchased from the American Society of Safety Engineers, 1800 E. Oakton Street, Des Plaines, IL 60018, (847) 699-2929.

Complete Confined Spaces Handbook, John F. Rekus, ISBN 1-87371-487-3, 1994, 400 pages.

The following publications are available from American Conference of Governmental Industrial Hygienists Publications, 1330 Kemper Meadow Drive, Cincinnati, OH 45240, (513) 742-2020.

Air Monitoring for Toxic Exposures, Shirley A. Ness, ISBN 0-442-20639-9, 1991, 534 pages.

Air Monitoring Instrumentation, Carol J. Maslansky & Steven P. Maslansky, ISBN 0-442-00973-9, 1993, 304 pages.

Industrial Ventilation: A Manual of Recommended Practice, 22nd ed., ISBN 1-882417-09-7, 1995, 470 pages.

1997 TLVs and BEIs, ISBN 1-882417-19-4, 1997, 156 pages.

The following are available from the American National Standards Institute (ANSI), 11 West 42nd Street, New York, NY 10036, (212) 642-4900.

Safety Requirements for Confined Spaces, ANSI Z117.1-1995.

Safety Requirements for Personal Fall Arrest Systems, Subsystems and Components, ANSI Z359.1-1992.

Requirements for Safety Felts, Harnesses, Lanyards and Lifelines for Construction and Demolition Use, ANSI A10.14-1991.

American National Standard for Respiratory Protection, ANSI Z88.2-1992.

CHAPTER 12

Lockout/Tagout

Employers are required to develop, document and utilize an energy control procedures program to control potentially hazardous energy.

The energy control procedures must:

- *specifically outline the scope, purpose, authorization, rules, and techniques to be utilized for the control of hazardous energy; and*
- *the means to enforce compliance including, but not limited to, the following:*
  - *—a specific statement of the intended use of the procedure;*
  - *—specific procedural steps for shutting down, isolating, blocking and securing machines and equipment to control hazardous energy;*
  - *—specific procedural steps for the placement, removal and transfer of lockout devices or tagout devices and the responsibility for them;*
  - *—specific requirements for testing a machine or equipment to determine and verify the effectiveness of lockout devices, tagout devices, and other energy control measures. (29 CFR 1910.147)*

12.1 INTRODUCTION

ONE of the major requirements in OSHA's Confined Space Entry Standard is to ensure that the hazards within a permit-required confined space have been removed or isolated. Removing the hazard is always the best way in which to protect entrants—however, removing all the hazards is, in many cases, impossible. Thus, OSHA requires the control of hazardous energy using isolation, blanking or blinding, disconnection, and/or lockout/tagout procedures.

Safety professionals know through experience that many workers mistake the results of atmospheric testing that show no hazard exists in a particular confined space as meaning that the space is totally safe for entry. Indeed, this might be the case—however, many other dangers inherent to confined spaces make entry into them hazardous. For example, if the confined space has some type of open liquid stream flowing through it, the chance for engulfment exists. If the space has electrical devices and circuitry inside, an electrocution hazard exists. If hazardous chemicals are stored and taken into the space, the potential for a hazardous atmosphere exists. Many confined spaces contain physical hazards, including piping and other obstructions—for example, rotating machinery is often housed within confined spaces.

To ensure that the confined space is indeed safe, any and all sources of hazardous energy must be isolated before entry is made. The primary method employed to accomplish this is through lockout/tagout procedures. In this chapter, I will first define the key terms associated with lockout/tagout and then present a sample of a written lockout/tagout program. As with all sample written programs presented in this text, this written lockout/tagout program has a huge advantage over many other such programs. It has been used in the real world—it has been tested.

12.2 LOCKOUT/TAGOUT KEY DEFINITIONS

- *Affected employee* An employee whose job requires him/her to operate or use a machine or equipment on which servicing or maintenance is being performed under lockout or tagout, or whose job requires him/her to work in an area in which such servicing or maintenance is being performed.
- *Authorized employee* A person who locks out or tags out machines or equipment to perform servicing or maintenance on them. An affected employee becomes an authorized employee when that employee's duties include performing servicing or maintenance covered under the organization's lockout/tagout program.
- *Capable of being locked out* An energy isolating device is capable of being locked out if it has a hasp or other means of attachment to which (or through which) a lock can be affixed or it has a locking mechanism built into it. Other energy isolating devices are capable of being locked out if lockout can be achieved without the need to dismantle, rebuild, or replace the energy isolating device or permanently alter its energy control capability.

  Note: After January 2, 1990, whenever replacement or major repair, renovation, or modification of a machine or equipment is performed, and whenever new machines or equipment are installed, energy isolating devices for such machines or equipment shall be designed to accept a lockout device.
- *Energized* Connected to an energy source or containing residual or stored energy.

- *Energy isolating device* A mechanical device that physically prevents the transmission or release of energy, including (but not limited to) the following: A manually operated electrical circuit breaker, a disconnect switch, a manually operated switch by which the conductors of a circuit can be disconnected from all ungrounded supply conductors, and, in addition, in which no pole can be operated independently; a line valve; a block; and any similar device used to block or isolate energy. Push buttons, selector switches, and other control circuit-type devices are not energy isolating devices.
- *Energy source* Any source of electrical, mechanical, hydraulic, pneumatic, chemical, thermal, or other energy.
- *Hot tap* A procedure used in the repair maintenance and service activities that involves welding on a piece of equipment (pipelines, vessels, or tanks) under pressure, to install connections or appurtenances. It is commonly used to replace or add sections of pipeline without the interruption of service for air, gas, water, steam, and petrochemical distribution systems.
- *Lockout* The placement of a lockout device on an energy isolating device, in accordance with an established procedure, ensuring that the energy isolating device and the equipment being controlled cannot be operated until the lockout device is removed.
- *Lockout device* A device that utilizes a positive means (such as a lock, either key or combination type) to hold an energy isolating device in the safe position and prevent the energizing of a machine or equipment. Included are blank flanges and bolted slip blinds.
- *Normal production operation* The utilization of a machine or equipment to perform its intended production function.
- *Selecting and/or maintenance* Work place activities such as constructing, installing, setting up, adjusting, inspecting, modifying, and maintaining and/or servicing machines or equipment. These activities include lubrication, cleaning, or unjamming of machines or equipment, and making adjustments or tool changes, where the employee may be exposed to the unexpected energization or start-up of the equipment or release of hazardous energy.
- *Setting up* Any work performed to prepare a machine or equipment to perform its normal production operation.
- *Tagout* The placement of a tagout device on an energy isolating device, in accordance with an established procedure, to indicate that the energy isolating device and the equipment being controlled may not be operated until the tagout device is removed.

- *Tagout device* A prominent warning device, such as a tag and a means of attachment, which can be securely fastened to an energy isolating device in accordance with an established procedure, to indicate that the energy isolating device and the equipment being controlled may not be operated until the tagout device is removed.

12.3 LOCKOUT/TAGOUT PROGRAM (A SAMPLE)

I. PURPOSE

This program has been developed to ensure protection of employees and to maintain compliance with OSHA Standard 1910.147, Control of Hazardous Energy.

These instructions establish the minimum requirements for the lockout or tagout of energy isolating devices whenever maintenance or servicing is done on machines or equipment. It shall be used to ensure that the machine or equipment is stopped, isolated from all potentially hazardous energy sources, and locked out before employees perform any servicing or maintenance. These procedures apply

(1) Whenever an employee has to remove or bypass a guard or other safety device
(2) Where such servicing results in the employee placing all or part of his or her body in a danger zone
(3) Where the unexpected energizing or start-up of the machine or equipment or release of stored energy could cause injury or death

II. AUTHORIZED EMPLOYEES

Authorized organization employees shall be trained in lockout/tagout procedure. Retraining will be done whenever an authorized employee's job responsibility changes or when the periodic inspection identifies procedural deficiencies. All affected (and other) employees whose work operations are or may be in the area shall be instructed in the purpose of use of lockout/tagout procedure (see Appendix A for list of authorized employees).

III. APPLICATION AND COMPLIANCE

This procedure applies to all organization employees and shall be enforced. All employees must signify that they have read and understand each part of this procedure by signing the training record (see Section IX of this procedure).

All employees are required to comply with the restrictions and limitations imposed upon them during the use of lockout or tagout. The employee, upon observing a machine that is locked or tagged out for servicing, *shall not* attempt to start or use that machine.

IV. TAGS

Tags are used to prevent the unexpected energization of equipment and generally are restricted to those controls that cannot be locked or safeguarded by any other method (see Figure 12.1).

(1) When the energy isolating device(s) cannot be locked out, use a tag to prevent inadvertent actuation.
(2) Tags will be attached to equipment or machinery, at the control panel, and at other points of operation.
(3) Tags must bear the words "DANGER," and "DO NOT OPERATE" or "DO NOT USE" printed on both sides of the tag.
(4) Tags shall bear the name, department, date, and telephone number of the employee (or department) performing the work.

V. LOCKS

Locks are employee-identifiable, key-operated padlocks. Lockout devices are those openings in equipment control handles of switches that can accept a padlock. Multiple lockout devices are those devices that have spaces for the application of more than one lock. This type of device is locked into the lockout opening, and subsequent locks prevent removal of the device (see Figure 12.1).

Authorized employees will be assigned their own individual lock with one unique key. A master key will generally not be available except under very unusual circumstances. A master key is necessary: It will be under the strict control of the work center supervisor. *Locks will not be removed* by anyone other than the authorized employee who placed it on the control, unless the procedure described in the "Special Condition" section of this program is followed. Employees working on the same machine will each use their own lock. When work continues from one shift to the next, employees leaving must remove their locks, and employees beginning work must apply their own locks.

VI. ENERGY ANALYSIS

Note: Example Statement (you should include such a statement in your own lockout/tagout program).

In 19 , organization's Safety Division surveyed each organization fa-

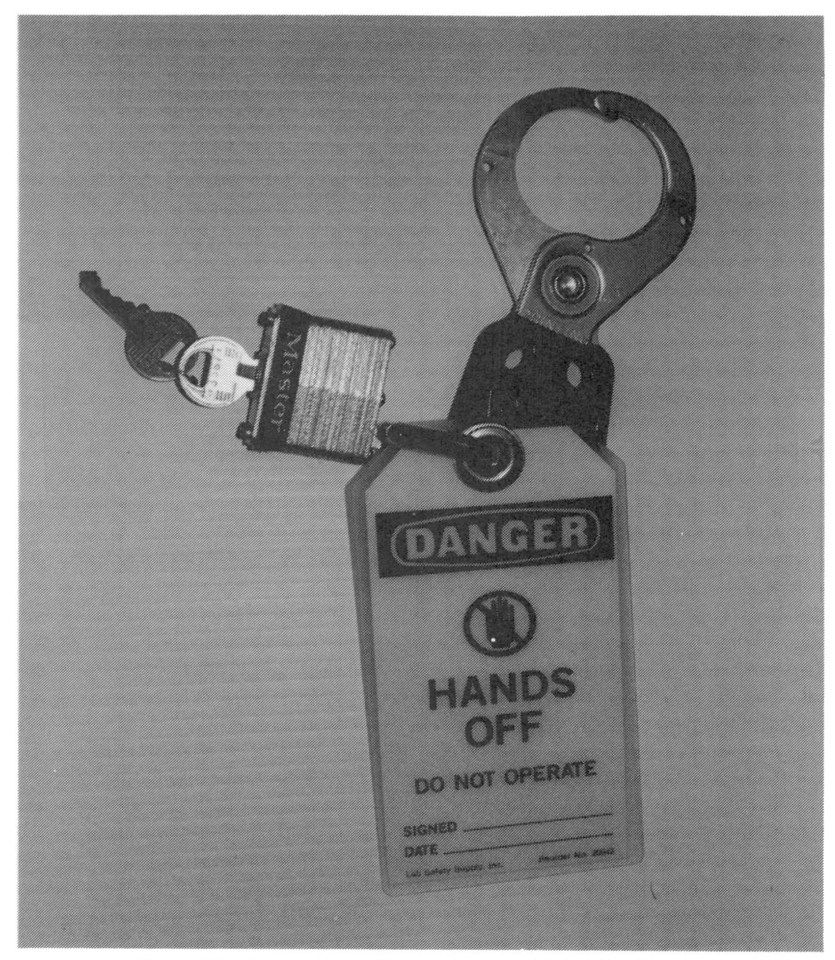

Figure 12.1 Tag, lock, and lockout hasp used for lockout/tagout.

cility and identified all machines or equipment covered by this program. All energy sources were identified along with their methods of control. Each time a new machine is installed, the organization Safety Division is responsible for performing an energy analysis. This energy analysis procedure is on-going and must be updated as required.

A. Equipment Energy Analysis Survey

Starting in June of 1989 and any time a new machine or new configuration to an existing machine has been installed since then, organization's

Safety Director conducted a survey of all organization's properties and made the following findings:

(List findings here; that is, list all equipment that must be locked/tagged out.)

Equipment ______________________

Equipment ______________________

Equipment ______________________

Etc. ____________________________

—and various equipment/machinery identified in organization's Maintenance Management System (preventive maintenance program) that have equipment-specific lockout/tagout procedures (see Appendix B). Organization is not required to document the equipment-specific procedure for each particular machine or equipment, because

(1) These machines and/or equipment were found to have no potential for stored or residual energy or re-accumulation of stored energy after shut down, which could endanger employees.
(2) The machine or equipment has a single energy source that can be readily identified and isolated.
(3) The isolation and locking out of that energy source will completely de-energize and deactivate the machine or equipment.
(4) The machine or equipment is isolated from the energy source and locked out during servicing or maintenance.
(5) A single lockout device will achieve a locked-out condition.
(6) The lockout device is under the exclusive control of the authorized employee performing the servicing or maintenance.
(7) The service or maintenance does not create hazards for other employees.
(8) Organization, in utilizing this exception, has had no accidents involving the unexpected activation or re-energization of the machine or equipment during servicing or maintenance.

For those machines and/or equipment that do not require equipment-specific lockout/tagout procedures, the lockout/tagout procedure in Section VII is to be used at organization.

VII. LOCKOUT/TAGOUT PROCEDURE

Lockout/tagout procedures for organization equipment other than equipment-specific lockout procedures for the incinerators and other covered equipment (see Appendix B) are

(1) Notify appropriate operations and maintenance supervisors of lockout/tagout.

(2) Place the main switch, valve, control, or operating lever in the "off," "closed," or "safe" position.

(3) *Check* and *test* to *make certain* that the proper controls have been identified and deactivated.

(4) Place a lock to secure the disconnection whenever possible. If a lock cannot be used on electrical equipment, an electrician shall remove fuses or disconnect circuit.

(5) If a system cannot be locked out with a lock, attach a "HOLD-OFF" or "DO NOT ENERGIZE" tag or other such tag to the switch, valve, or lever. If the organization work center does not use employee-identifiable locks, a lock and tag must be used together.

(6) When auxiliary equipment or machine controls are powered by separate supply sources, such equipment or controls shall be locked or tagged to prevent any hazard that may be caused by operating the equipment or exposure to live circuits.

(7) When equipment uses pneumatic or hydraulic power, pressure in lines or accumulators shall be checked. Using whatever safe means possible, this pressure shall be disconnected or pressure lines disconnected.

(8) When stored energy is a factor as a result of position, spring tension, or counterweighting, the equipment shall be placed in the bottom or closed position, or it shall be blocked to prevent movement.

(9) When the work involves more than one person, additional employees shall attach their locks and tags as they report.

(10) When outside contractors are involved, the equipment shall be locked out and tagged in accordance with this procedure by the project manager supervising the work. Only in emergency cases is equipment to be shut down by other than an organization representative.

(11) When the servicing or maintenance is completed and the machine or equipment is ready to return to normal operating condition, the following steps will be used:

a. Check the machine or equipment and the immediate area around the machine to ensure that tools, materials, and other nonessential items have been removed and that the machine or equipment components are operationally intact. Ensure that all guards have been replaced.

b. Check the work area to ensure that all employees have been safely positioned or removed from the area.

c. Verify that the controls are in neutral.

d. Remove the lockout devices and re-energize the machine or equipment. Each lockout/tagout device will be removed from each energy isolating device by the employee who applied the device.
e. Notify affected employees that the servicing or maintenance is completed, and that the machine or equipment is again ready for use.

VIII. PERIODIC INSPECTIONS

A periodic inspection of the energy control procedure(s) will be conducted at least quarterly to ensure that the requirements of the program and the standard are being followed to ensure full employee protection.

The periodic inspection will be performed by the work center supervisor and/or Safety Division personnel.

The inspection will be conducted to identify any program inadequacies that need correcting. The inspection will review the employee's responsibilities under the procedure being inspected.

Organization will provide certification that the inspection of a lockout/tagout has been performed. The certification will identify the machine on which the lockout/tagout is being utilized, the date of the inspection, names of employees involved, and the name of the individual performing the inspection. These entries will be made on the work center quarterly inspection form and the periodic inspection worksheet shown in Figure 12.2. These forms are to be kept on file by the Safety Division for 1 year.

IX. TRAINING

Training will be provided by organization's Safety Division, to ensure that the purpose and function of the lockout/tagout program are understood by all employees and that the knowledge and skills required for safe application, usage, and removal of energy controls are acquired by employees. The training shall include the following:

(1) Each AUTHORIZED (person who actually locks/tags out) employee shall receive training in the recognition of hazardous energy sources, the type and magnitude of the energy available in the work place, and the methods and means necessary for energy isolation and control.
(2) Each AFFECTED (person affected by the lockout/tagout) employee shall be instructed in the purpose and use of the energy control procedure.

PERIODIC INSPECTION WORKSHEET
(CERTIFICATION SHEET)

Lockout/tagout program inspection for organization work centers.

Name of authorized employee ______________________

DEPARTMENT	MACHINERY	TYPE OF POWER	METHOD OF CONTROL
____________	____________	Electrical	__________
		Mechanical	__________
		Hydraulic	__________
		Pneumatic	__________
		Chemical	__________

Did the energy control procedures identify all hazards associated with this machine? ______________________

If not, describe deficiencies. ______________________

____

Is retraining required? ______________________

Has retraining been done? (DATE) ______________________

______________________ ______________________

Signature of Inspector Date

Figure 12.2

(3) All OTHER employees whose work operations are or may be in an area where energy control procedures may be used shall be instructed about the procedure and about the prohibition relating to attempts to restart or re-energize machines or equipment that are locked out or tagged out.

When tagout is used, employees shall receive additional training on the limitations of tags.

A training record (see Figure 12.3) will be used to record each employee's training.

X. SPECIAL CONDITIONS

Lockout/tagout removal when authorized employee is absent—When the authorized employee who applied the lockout or tagout device is not available to remove it, that device may be removed under the direction of the supervisor, provided that specific procedures and training for such removal have been developed, documented, and incorporated into the lockout/tagout program. Specific procedures include the following elements:

TRAINING RECORD
(SAMPLE OF)

TRAINING ROSTER

LOCKOUT/TAGOUT PROGRAM

Date: ____________

In accordance with the recording and training requirements of the Lockout/Tagout Program, OSHA 29 CFR 1910.147, I have received Safety training in Lockout/Tagout requirements and procedures. I have agreed to verify my understanding and training of 1910.147 by signing and dating this form. This training meets requirements of 29 CFR 1910.147.

NAME	WORK CENTER
______________________	______________________

Figure 12.3

LOCKOUT/TAGOUT REMOVAL
WHEN AUTHORIZED EMPLOYEE IS ABSENT

Department ________________ Machine/Equipment ________________ Lock ID # ________________

Authorized Employee ______________________ Type of Work Being Done ______________________

Verification that employee is not on site (Yes) ______ (No) ______

NOTE: ____
____

Verification that employee has been informed before coming back to work that his/her lockout/tagout device has been removed.

(Yes) ________ (No) ______

NOTE: ____
____

______________________ ______________________
Supervisor Date

Figure 12.4

(1) Verifying that the authorized employee who applied the device is not at the facility
(2) Making all reasonable efforts to contact the authorized employee to inform him/her that his/her lockout or tagout device has been removed
(3) Ensuring that the authorized employee has this knowledge before he/she resumes work at that facility
(4) Completing the "Lockout/Tagout Removal When Authorized Employee Is Absent" form—the form shall be kept by the work center supervisor (see Figure 12.4)

XI. METHODS OF INFORMING OUTSIDE CONTRACTORS OF PROCEDURES

Whenever outside servicing personnel are to be engaged in lockout/tagout activities covered by this program, organization AND the outside employer *will inform each other* of their respective lockout or tagout procedures.

Organization employees will be trained to understand and comply with the restrictions and prohibitions of the outside contractor's lockout/tagout energy control program.

The Safety Division will provide to the organization work center supervisor a contractor briefing form (Figure 12.5) that will inform the contractor on precautionary measures involved in performing lockout/tagout on the site.

RECORD OF NOTIFICATION

ORGANIZATION AND OUTSIDE CONTRACTOR SAFETY ARRIVAL CONFERENCE

In accordance with the recordkeeping and training requirements under OSHA 1910.119 and other applicable standards, I have received a safety brief from organization Safety Division/Plant personnel covering organization's lockout/tagout program. I further understand that organization expects all outside contractors, including subcontractors, to perform construction activities under OSHA-required guidelines and organization procedures. It is understood that all information regarding the hazards, safety procedures, lockout/tagout procedures shall be disseminated to all people, sub-contractors, and agents employed either directly or indirectly by us. Regarding lockout/tagout procedures, organization and the contractor must discuss these operations and work in a coordinated effort so as to prevent injury to anyone.

Signature	Printed Name	Company Name
________________	________________	________________
________________	________________	________________
________________	________________	________________

Figure 12.5

APPENDIX A

List of Authorized Employees

Alpha Department
Plant Manager's/Superintendents
Chief Operators
Lead Operators
STPOs/STMOs
SHPOs/SHMOs

Bravo Department
Electrical Superintendent
Electricians/Instrument Specialists
Lead Mechanic
Machinists
Carpenters

Charlie Department
System Superintendents
System Chief Foremen
Systems Foreman
Pump Station Supervisors
Pump Station Checkers
Interceptor Supervisor
Interceptor Technicians
Interceptor Assistants
Maintenance Crew Members
Reliability Technicians
Reliability Manager
Reliability Supervisor
Chemical Technician/Assistant
Qualified Helpers

Delta Department
Industrial Waste Managers
Industrial Waste Technicians
Industrial Waste Specialists
Technical Service Division Techs/Specs
Stormwater Specialist/Assistants
Environmental Scientists

APPENDIX B

Equipment-Specific Lockout/Tagout Procedures Specified Equipment

Equipment ____________________

Equipment ____________________

Equipment ____________________

Etc. ________________________

____

12.4 REFERENCE

The Office of the Federal Register. *Code of Federal Regulations Title 29 Parts 1900–1910 (.147)*, Office of the Federal Register, Washington, D.C., 1995.

CHAPTER 13

Respiratory Protection

Written procedures shall be prepared covering safe use of respirators in dangerous atmospheres that might be encountered in normal operations or in emergencies. Personnel shall be familiar with these procedures and the available respirators. [29 CFR 1910.134 (c)]

13.1 INTRODUCTION

RESPIRATORS are devices that can allow workers to safely breathe without inhaling particles or toxic gases. Two basic types (1) *air-purifying,* which filters dangerous substances from the air, and (2) *air-supplying,* which delivers a supply of safe breathing air from a tank (SCBA), a group of tanks (cascade system), or an uncontaminated area nearby via a hose or airline to the mask, are in common use.

For permit-required confined space entry operations, respiratory protection is safety equipment that is always required for entry into an IDLH space or must be readily available for emergency use and rescue if conditions change in a non-IDLH space. Note, however, that *only air-supplying respirators should be used in confined spaces where there is not enough oxygen.*

Selecting the proper respirator for the job, the hazard, and the worker is very important, as is thorough training in the use and limitations of respirators. Compliance with OSHA's Respiratory Standard and ensuring safe confined space entry begins with having written procedures covering all applicable aspects of respiratory protection. This chapter presents a sample written respiratory protection program; again, this sample program has been used for several years and has proven its worth and effectiveness.

13.2 WRITTEN RESPIRATORY PROTECTION PROGRAM (A SAMPLE)

I. INTRODUCTION

The Occupational Safety and Health Act (OSH Act) requires that every employer provide a safe and healthful work environment. This includes ensuring workers are protected from unacceptable levels of airborne hazards. While most air is safe to breathe, certain work operations and locations have characteristic problems of air contamination. Control measures are required to reduce airborne hazard concentrations to safe levels. When controls are not feasible, or while they are being implemented, workers must wear approved respiratory protection.

The organization has adopted this respiratory protection program to comply with OSHA regulations (as set forth in 29 CFR 1910.134) and to do all that is possible to protect those employees who are filling a job classification that requires respirator use in the performance of their duties. All departments and work centers are included and must adhere to the requirements set forth in this program. Organization's respiratory protection program is an organized approach for ensuring employees a safe work place by providing specific requirements in these areas:

(1) Designation of individual departmental responsibilities
(2) Definition of various terms used in the respiratory protection program
(3) Designation of types of respirators and their applications
(4) Designation of procedures for respirator selection and distribution
(5) Designation to procedures to be used for inspection and maintenance of respirators
(6) Designation of procedures for employee respirator fit-testing
(7) Designation of a procedure for medical surveillance
(8) Designation of a training program for personnel participating in organization's respiratory protection program
(9) Documentation procedure for personnel participating in organization's respiratory protection program

II. RESPONSIBILITIES

Department directors will be responsible for the following:

(1) Implement and ensure compliance of departmental personnel with organization's respiratory protection program

(2) Specify the job classifications that use respirators, and ensure this job requirement is included in job descriptions for these classifications

Organization's Safety Division has the following responsibilities under organization's respiratory protection program:

(1) Develop and modify as necessary organization's written respiratory protection program
(2) Check and review quarterly all work center programs, including work center respirator inspection record
(3) Compile and maintain a master respirator inventory list for organization
(4) Implement an on-going respirator training program
(5) Conduct initial and annual employee fit testing
(6) Provide initial and annual spirometric evaluation to ensure that employees are capable of wearing a respirator under their given work conditions
(7) Provide technical assistance in determining the need for respirators and in the selection of appropriate types of respirators
(8) Forward training, fit test, initial/annual spirometric evaluation, and medical doctor's evaluation for suitability to wear a respirator to Human Resources Manager for inclusion into employee's personnel record
(9) Inspect quarterly the accuracy and proper maintenance of records specified in this program
(10) Conduct air-quality tests annually on internal combustion engine-driven air-line respirator compressors to ensure proper air quality

Organization supervisory personnel are responsible for the following:

(1) Ensure that respirators are available to employees as needed
(2) Ensure that employees wear appropriate respirators as required
(3) Ensure inspection of cartridge type respirators on a monthly basis, and self-contained breathing apparatus (SCBAs) and Air-Line Hose Mask systems on a weekly and monthly basis. Ensure records of respirator inspections are maintained
(4) Ensure employees are fit tested and receive initial/annual spirometric evaluation prior to using a respirator

The employee is responsible for the following:

(1) Use supplied respirators in accordance with instructions and training
(2) Clean, disinfect, inspect, and store assigned respirator(s) properly
(3) Perform self fit test prior to each use and ensure that manageable

physical obstructions such as facial hair do not interfere with respirator fit

(4) Report respirator malfunctions to supervision and conduct "After Use Inspection" of SCBA type respirator
(5) Report any poor health conditions that may preclude safe respirator usage

Organization Human Resources Manager is responsible for the following:

(1) Schedule required initial medical examination and spirometric evaluation for all new employees who fill a job classification requiring the use of respirators
(2) Maintain records of employee medical, spirometric, and fit test results

III. DEFINITION OF TERMS

Organization's respiratory protection program defines various terms as follows:

- *Aerosol* A suspension of solid particles or liquid droplets in a gaseous medium.
- *Asbestos* A broad mineralogical term applied to numerous fibrous silicates composted of silicon, oxygen, hydrogen, and metallic ions, like sodium, magnesium, calcium, and iron. At least six forms of asbestos occur naturally. Types of asbestos that are currently regulated—actinolite, amosite, anthophylite, chrysotile, crocidolite, and tremolite.
- *Banana oil* A liquid that has a strong smell of bananas, used to check for general sealing of a respirator during fit-testing.
- *Blasting abrasive* A chemical contaminant composed of silica, silicates, carbonates, lead, cadmium, or zinc and is classified as a dust.
- *Breathing resistance* The resistance that can build up in a chemical respirator cartridge that has become clogged by particulates.
- *Chemical hazard* Any chemical that has the capacity to produce injury or illness when taken into the body.
- *Cleaning respirators* Cleaning respirators involves the use of a mild detergent and rinsing with potable water.
- *Dust* A dispersion of tiny solid airborne particles produced by grinding or crushing operations.
- *Fit-testing* An evaluation of the ability of a respiratory device to interface with the wearer in such a manner as to pre-

vent the work place atmosphere from entering the worker's respiratory system.

- *Forced Expiratory Volume (FEV1)* That volume of air that can be forcibly expelled during the first second of expiration.
- *Forced Vital Capacity (FVC)* The maximal volume of air that can be exhaled forcefully after a maximal inhalation.
- *Fume* Solid particles generated by condensation from the gaseous state.
- *Gas* A substance that is in the gaseous state at ordinary temperature and pressure.
- *IDLH (immediately dangerous to life and health)* Any condition that poses an immediate threat to life or that is likely to result in acute or immediately severe health effects.
- *Irritant smoke (stannic oxychloride)* A chemical used to check for general sealing of a respirator during a fit test.
- *Mist* A dispersion of liquid particulates.
- *Oxygen deficiency* Any level below the PEL of 19.5 percent.
- *Particulates* Dusts, mists, and fumes.
- *Permissible Exposure Limit (PEL)* The maximum time weighted average concentration of a substance in air that a person can be exposed to during an 8-hour shift.

PEL/IDHL Chart

Chemical Name	PEL (8-hr. average)	IDLH
1. Ammonia	50 ppm	300 ppm
2. Carbon dioxide	5,000 ppm	50,000 ppm
3. Carbon monoxide	50 ppm	1,200 ppm
4. Sodium hydroxide	Must use SCBA to enter	
5. Sulfur dioxide	5 ppm	100 ppm
6. Chlorine	1 ppm	10 ppm
7. Hydrogen chloride	5 ppm	100 ppm
8. Hydrogen sulfide	10 ppm	100 ppm
9. Propane	1,000 ppm	2,100 ppm
10. Oxygen	19.5% (min)	—
11. Flammable	10% LEL	

- *Respirator* A face mask that filters out harmful gases and particles from air, enabling a person to breathe and work safely.
- *Respiratory hazard* Any hazard that enters the human body by inhalation.

- *Saccharin* A chemical that is sometimes used to check for general sealing of a respirator during fit-testing.
- *Smoke* Particles that result from incomplete combustion.
- *Spirometric evaluation* A test used to measure pulmonary function. A measurement of FVC and FEV1 of 70 percent or greater is satisfactory. A measurement of less than 70 percent may require further pulmonary function evaluation by a medical doctor.
- *Vapor* The gaseous state of a substance that is liquid or solid at ordinary temperature and pressure.

IV. TYPES OF RESPIRATORS

A. Chemical Cartridge Respirators

(1) Description: Chemical cartridge respirators may be considered to be low-capacity gas masks. They consist of a facepiece, which fits over the nose and mouth of the wearer. Attached directly to the facepiece is a small replaceable filter-chemical cartridge.

(2) Application: Usually this type of respiratory protection equipment is used where there is exposure to solvent vapors or dust and particulate matter as with sandblasting, spray coating, or degreasing. They may not be worn in IDLH atmospheres.

B. Air-Line Respirators (Helmet, Hoods, and Masks)—Cascade-Fed or Compressor-Fed

(1) Description: These devices provide air to the wearer through a small-diameter, high-pressure hose line from a source of uncontaminated air. The source is usually derived from a compressed air line with a valve in the hose to reduce the pressure. A filter must be included in the hose line (between the compressed air line and the respirator) to remove oil and water mists, oil vapors, and any particulate matter that may be present in the compressed air. Internally lubricated compressors require that precautions be taken against overheating, because the heated oil will break down and form carbon monoxide. Where the air supply for air-line respirators is taken from the compressed air line, a carbon monoxide alarm must be installed in the air supply system. Completion of prior-to-operation preventive maintenance check of the carbon monoxide alarm system is critical.

(2) Application: Air-line respirators used in industrial application for confined space entry (IDLH atmosphere) must be equipped with an emergency escape bottle.

C. Self-Contained Breathing Apparatus (SCBA)

(1) Description: This type of respirator provides Grade D breathing air (not pure oxygen), either from compressed air or breathing air cylinders or by chemical action in the canister attached to the apparatus. It enables the wearer to be independent of any outside source of air. This equipment may be operable for periods between 30 minutes and 2 hours. The operation of the self-contained breathing apparatus is fairly complex, and it is, therefore, necessary that the wearer have special training before being permitted to use it in an emergency situation.

(2) Application: Because the oxygen-producing mechanism is self-contained in the apparatus, it is the only type of equipment that provides complete protection and at the same time permits the wearer to travel for considerable distances from a source of respirable air. SCBAs (with the exception of hot work activities) can be utilized in many industrial applications.

V. RESPIRATOR SELECTION AND DISTRIBUTION PROCEDURES

Respirators are selected by work center supervisors. See Figures 13.1 and 13.2 for selection process assistance. Figure 13.1 is a flow chart that provides guidelines for selection of an appropriate respirator. For cartridge-type respirators, Figure 13.2 provides a listing of the appropriate cartridges to be used with various atmospheric contaminants. Selection is based on matching the proper color-coded cartridge with the type of protection desired. Selection is also dependent upon the quality of fit and the nature of the work being done. Cartridge-type respirators are issued to the individuals who are required to use them. Each individually assigned respirator is identified in a way that does not interfere with its performance. Questions about the selection process are to be referred to the Safety Division.

VI. RESPIRATOR INSPECTION, MAINTENANCE, CLEANING, AND STORAGE

To retain their original effectiveness, respirators should be periodically inspected, maintained, cleaned, and properly stored.

Note: In the following sections, several references are made to various inspection records. You should design site-specific standard record forms and inspection records for use with your respiratory protection program.

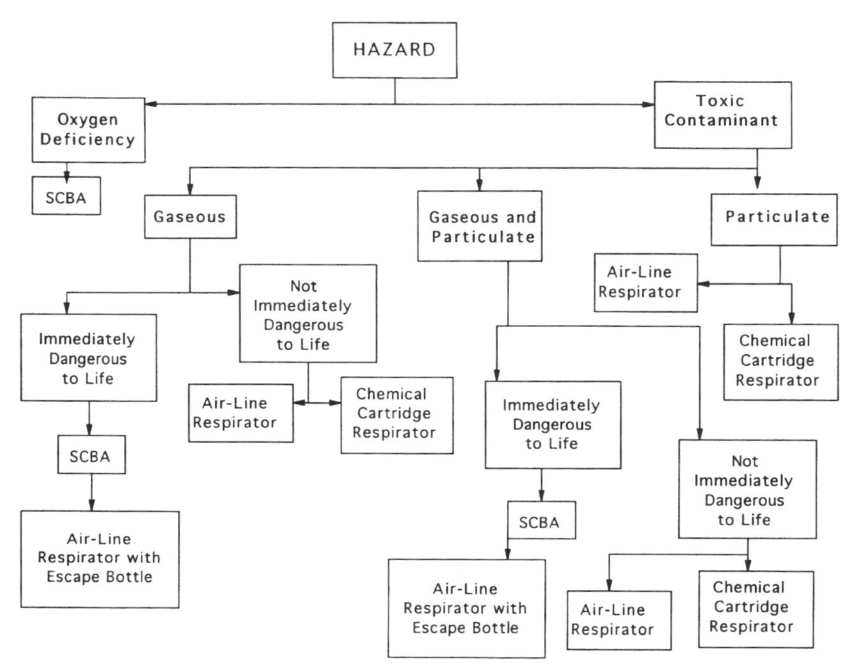

Figure 13.1 Respirator selection flow chart.

Atmospheric Contaminants to Be Protected Against	Color assigned
Acid Gases	White
Chlorine	White/Yellow Stripe
Organic Vapors	Black
Ammonia Gas	Green
Acid Gas and Organic Vapors	Green/White Stripe
Carbon Monoxide	Blue
Acid, Gases, Organic Vapors, and Ammonia Gases	Brown
Particulates (dust, fumes, mists, fog, or smoke)	1/2″ Grey Stripe around Normal canister color
All of the Above Atmospheric Contaminants	Red with 1.5″ Grey Stripe
Asbestos	HEPA (Pink)

Figure 13.2 Respirator cartridge color-coding system.

A. Inspection

(1) Respirators should be inspected before and after each use, after cleaning, and whenever cartridges or cylinders are changed. Appropriate entries should be made in a respirator "Inspection after Each Use" record.
(2) If a half face air purifying respirator is taken out of use, indicate it on the inspection records. The respirator must be inspected thoroughly before it is put back in use.
(3) Work center supervisors shall ensure all cartridge-type respirators are inspected once per month and make appropriate entries in a "Supervisor's Monthly Respirator Inspection Checklist." The work center supervisor or designated person shall inspect all SCBAs and air-line respirators weekly and monthly, and make appropriate entries in a "SCBA/Air-Line Respirator Weekly and Monthly Inspection and Maintenance Checklist" record. These records are to be kept by each work center for a period of 3 years.
(4) Safety Division personnel will inspect these records quarterly.

B. Maintenance

Respirators that do not pass inspection must be replaced or repaired prior to use. Respirator repairs are limited to the changing of canisters, cartridges, cylinders, filters, head straps, and other items as recommended by the manufacturer. No attempt should be made to replace components or make adjustments, modifications, or repairs beyond the manufacturer's recommendations.

C. Cleaning

Individually assigned cartridge respirators are cleaned as frequently as necessary by the assignee to ensure proper protection is provided. SCBA respirators are cleaned after each use.

The following procedure is used for cleaning respirators:

(1) Filters, cartridges, or canisters are removed before washing the respirator and are discarded and replaced as necessary.
(2) Cartridge-type and SCBA respirator facepieces are washed in a detergent solution, rinsed in clean potable water, and allowed to dry in a clean area. A clean brush is used to scrub the respirator to remove adhering dirt.

D. Storage

After inspection, cleaning, and necessary repairs, respirators are stored to protect against dust, sunlight, heat, extreme heat, extreme cold, excessive moisture, or damaging chemicals. Respirators are to be stored in plastic bags or the original case. Individuals who are assigned respirators are to store their respirators in assigned personal lockers. General use SCBAs are to be stored in designated cabinets, racks, or lockers with other protective equipment. Respirators are not to be stored in tool boxes or in the open. Individual cartridges or masks with cartridges are to be sealed in plastic bags to preserve their effectiveness.

VII. RESPIRATOR FIT-TESTING

The respiratory protection program provides standards for respirator fit-testing. The goal of respirator fit-testing is to (1) provide the employee with a face seal on a respirator that exhibits the most protective and comfortable fit and (2) to instruct the employee on the proper use of respirators and their limitations.

There are three levels of fit-testing: initial, annual, and pre-use self-testing.

A. The Initial and Annual Fit Tests Are Rigorous Procedures Used to Determine Whether the Employee Can Safely Wear a Respirator

The initial and annual tests are conducted by the Safety Division. Both tests utilize the cartridge and SCBA-type respirator to check each employee's suitability for wearing either type. Fit-testing requires special equipment and test chemicals such as banana oil, irritant smoke, or saccharin (see Figure 13.3). In general, any changes to the face or mouth may alter respirator fit and may require the use of a specially fitted respirator. Organization's Safety-Division will make this determination. Upon completion of initial fit-testing, the safety division forwards the original of the employee's Fit Test Record to the Human Resources Manager for inclusion in the employee's file. A copy will be forwarded to the affected work center supervisor.

B. Pre-Use Self-Testing—A Routine Requirement for All Employees Who Wear Respirators

Each time the respirator is used, it must be checked for positive and negative seal. The Safety Division will train supervisors on this procedure. Supervisors are responsible for training employees in their individual work centers.

(1) Positive pressure check procedure (cartridge-style respirator): After

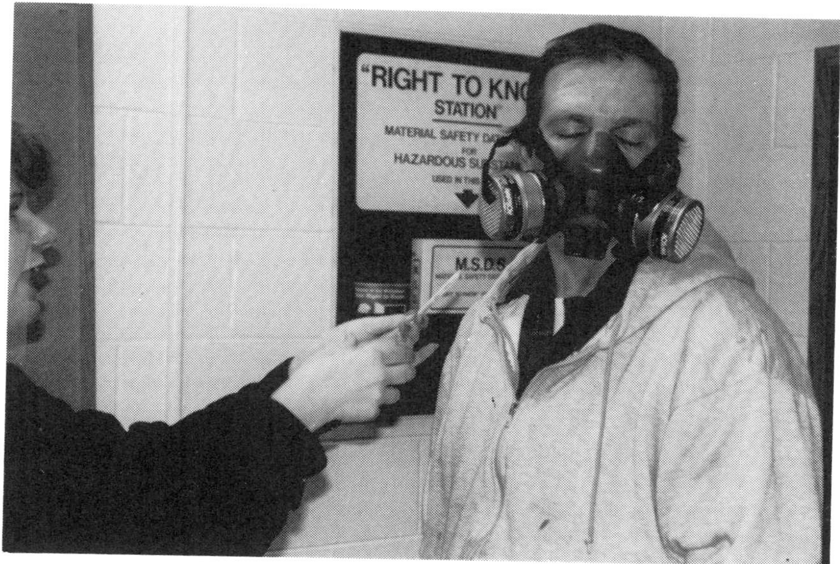

Figure 13.3 Fit testing worker for compliance with OSHA's respiratory protection standard.

the respirator has been put in place and straps have been adjusted for firm but comfortable tension, the exhalation valve is blocked by the wearer's palm. He or she takes a deep breath and gently exhales a *little* air. Hold the breath for 10 seconds. If the mask fits properly, it will feel as if it wants to pop away from the face, but no leakage will occur.

(2) Negative pressure check procedure (cartridge-style respirator): While still wearing the respirator, cover both filter cartridges with the palms, and inhale slightly to partially collapse the mask. Hold this negative pressure for 10 seconds. If no air leaks into the mask, it can be assumed the mask is fitting properly.

Note: Self-test fit-testing can be conducted, for both positive and negative pressure checks, on the SCBA-type respirator by crimping the hoses with fingers, vice blocking airways with palm of hands.

If either test shows leakage, the following procedure should be followed:

(1) Ensure mask is clean. A dirty or deteriorated mask will not seal properly, nor will one that has been stored in a distorted position. Proper cleaning and storage procedures must be used.
(2) Adjust the head straps to have snug, uniform tension on the mask. If

only extreme tension on the straps will seal the respirator, report this to the supervisor. Note that a mask with uncomfortably tight straps rapidly becomes obnoxious to the wearer.

Standard 1910.134 (g) (1) (A) states: Personnel with facial hair that comes between the sealing surface of the facepiece and the face or that interferes with valve function shall not be permitted to wear tight-fitting respirators. Thus, respirator wearers with beards or side burns that interfere with the face-seal are prohibited from wearing tight-fitting respirators on the job.

Dental changes—loss of teeth, new dentures, braces, and so forth—may affect respirator fit and may require a new fitting with a different type of mask.

Note: Any change to the face or mouth that may alter respirator fit must be brought to the immediate attention of the work center supervisor.

VIII. MEDICAL SURVEILLANCE

OSHA states that no one should be assigned a task requiring use of respirators unless they are found medically fit to wear a respirator by competent medical authorities. Organization's respiratory protection program will include a medical surveillance procedure that includes the following items.

A. Pre-Employment Physical/Spirometric Evaluation/5-Year Follow-Up Physical Exam

All new and regular employees who fill job classifications that require respirator use in the performance of their duties are required to pass an initial medical examination to determine fitness to wear respiratory protection on the job. Annual spirometric evaluations will be conducted to ensure that employees covered under this program meet the OSHA requirements for fitness to wear respirators. On a continuous 5-year basis, all organization employees covered under this program will be re-examined by competent medical authorities to ensure their continued fitness to wear respiratory protection on the job.

Each Department Director will specify which job classifications require the employee to use respirators. Pre-employment and 5-year follow-up medical evaluation will be conducted by a medical doctor. Spirometric evaluation will be conducted by the Safety Division. The Safety Division will forward the employee's spirometry results to the Human Resources Manager for inclusion in the employee's personnel file.

B. Annual Spirometric Evaluation

Annual spirometric evaluations will be conducted by the Safety Divi-

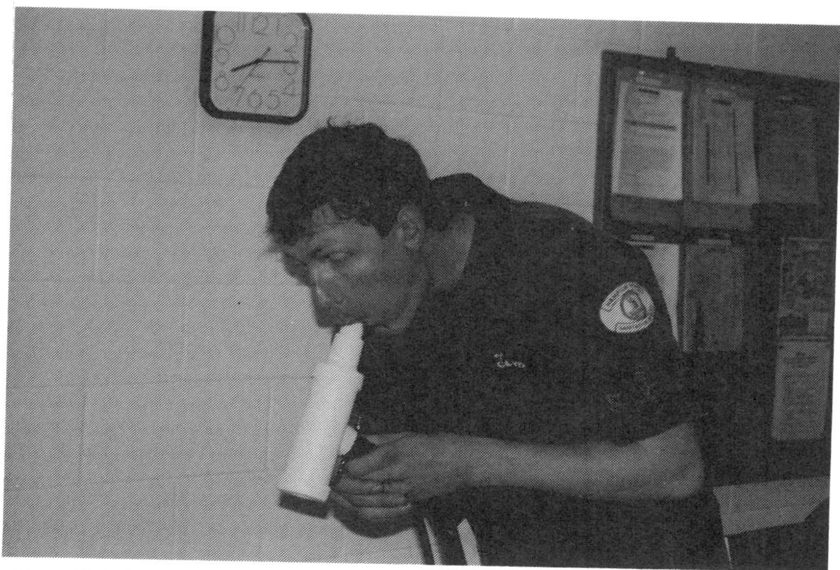

Figure 13.4 Spirometric examination to test pulmonary function and suitability to wear respirators.

sion on all employees filling job classifications requiring the use of respirators in the performance of their duties. Spirometry testing will be used to measure Forced Vital Capacity (FVC) and Forced Expiratory Volume-1 second (FEV1) (see Figure 13.4). If FVC is less than 75 percent and/or FEV1 is less than 70 percent, the employee will not be allowed to wear a respirator unless a written waiver is obtained from a medical doctor. The supervisor determines whether the employee can be exempted from work functions that require wearing a respirator.

Note: Organization will make reasonable accommodations to allow employees to retain their current positions with specified medical restrictions on respirator use.

The Safety Division will route annual results of spirometric testing to the Human Resources Manager for inclusion in each employee's personnel file and will notify appropriate supervisors of any employee who fails the test.

IX. TRAINING

No worker may wear a respirator before spirometric evaluation, medical evaluation, fit-testing, and training have all been completed and documented.

(1) The Safety Division holds the responsibility for providing employee respirator training.

(2) Supervisors are the day-to-day monitors of the program and have the responsibility to perform refresher training and to ensure self fit-testing is accomplished by their employees as needed.

Available dates for Safety Division-administered training sessions will be published on a routine basis. Supervisors are responsible for scheduling their new employees for the next available session. Training on respiratory protection is also conducted at New Employee Safety Orientation sessions.

This respiratory protection program is subject to changes and improvements as new regulations and technologies emerge. The Safety Division will train supervisors and employees as applicable on any new information.

X. DOCUMENTATION PROCEDURES

Documentation of safety training is very important. OSHA insists that certain records be maintained on all employees. All safety training records should be considered legal records; the likelihood of having to use safety training records in a court of law is real.

The following information will be maintained by the Safety Division:

(1) Date and location of initial employee training
(2) Inventory records of all organization respirators

The following information will be processed by the Human Resources Manager for inclusion in the employee's personnel file.

(1) Results of annual employee fit-testing
(2) Results of new employee medical evaluation and annual spirometric testing (to remain on file for 5 years)

Supervisors will maintain:

(1) A file of respirator inspection records
(2) Respirator inventory records

Note: The maintenance and accuracy of all records specified in this will be inspected quarterly by the Safety Division.

XI. PROCEDURE FOR SAFE USE OF SCBA/SUPPLIED AIR RESPIRATORS

To be in compliance with 1910.134 (e) (3), organization is providing these written procedures covering the safe use of respirators (SCBA and supplied air respirators only).

Note: Air purifying/chemical cartridge respirators are to be used only for coatings and sand blasting operations and *never* for confined space entry or any other activity where oxygen deficiency or atmospheric contaminants are present.

SCBAs and/or supplied air (with emergency escape bottles) are to be used in all situations that involve chemical handling, confined space entry during normal operations, and in emergencies.

A. Safe Use Procedure in Dangerous Atmospheres

This written procedure is prepared for safe respirator use in IDLH atmospheres that may occur in normal operations or emergencies.

All organization personnel (covered under this program) are to be familiar with these procedures and respirators.

(1) Inspect all respirator equipment prior to use to ensure that it is complete and in good repair.

(2) Ensure respirator facepiece is correct size for your face; perform a self fit-test.

(3) Ensure that available air is adequate for the expected time to be used.
Note: No organization employee should don an SCBA that is not 100 percent full.

(4) Test all alarms on the respirator to ensure that they work.

(5) At least two fully trained and certified standby/rescue persons, equipped with proper rescue equipment, including an SCBA, will be present in the nearest safe area for emergency rescue of those wearing respirators in an IDLH atmosphere.

(6) Communications (visual, voice, signal line, telephone, radio, or other suitable type) will be maintained among all people present (those in the IDLH atmosphere and the standby person or people). The respirator wearers are to be equipped with safety harness and safety lines to permit their removal from the IDLH atmosphere if they are overcome.

(7) The atmospheres in a confined space may be immediately dangerous to life or health (IDLH) because of toxic air contaminants or lack of oxygen. Before any organization employee enters a confined space, tests must be performed to determine the presence and concentration of any flammable vapor or gas, or any toxic airborne particulate, vapor, or gas, and to determine the oxygen concentration (follow all procedures as outlined in organization's confined space program).

(8) No one is to enter if a flammable substance exceeds the lower explosive limit (LEL). No one should enter without wearing the proper type of respirator if any air contaminant exceeds the established permissible exposure limit (PEL) or if there is an oxygen deficiency. Ensure that

the confined space is force-ventilated to keep the flammable substance at a safe level.

Note: Even if the contaminant concentration is below the established breathing time-weighted average (TWA) limit and there is enough oxygen, the safest procedure is to ventilate the entire space continuously and to monitor the contaminant and oxygen concentrations continuously if people are to work in the confined space without respirators.

(9) If the atmosphere in a confined space is IDLH owing to a high concentration of air contaminant or oxygen deficiency, those who must enter the space to perform work must wear a pressure-demand SCBA or a combination pressure-demand air-line and self-contained breathing apparatus that always maintains positive air pressure inside the respiratory inlet covering. Fully trained and equipped rescue crew must be on-site and ready to respond if needed. This is the best safety practice for confined space entry and *is required* at organization.

13.3 REFERENCE

The Office of the Federal Register. *Code of Federal Regulations Title 29 Parts 1900–1910 (.134)*, Office of the Federal Register, 1995.

CHAPTER 14

Hot Work Permits

The employer shall issue a hot work permit for hot work operations conducted on or near a covered process [including confined spaces].

The permit shall document that the fire prevention and protection requirements in 29 CFR § 1910.252 (a) [Fire Prevention and Protection] have been implemented prior to beginning the hot work operations; it shall indicate the date(s) authorized for hot works; and identify the object on which hot work is to be performed. The permit shall be kept on file until completion of the hot work operations. (29 CFR 1910.119, .134, .252, Code of Federal Regs., 1995)

14.1 INTRODUCTION

IN the two previous chapters, I have pointed to the importance of complying with OSHA's lockout/tagout and respiratory protection programs in conducting proper and safe confined space operations. This chapter finishes our discussion of confined space entry operations by discussing another OSHA requirement important to confined space entry: Hot work permits.

Under the General Industry Standard, OSHA specifies in its Process Safety Management Standard (29 CFR 1910.119), in its Confined Space Standard (29 CFR 1910.146), and in its Fire Prevention and Protection Standard (29 CFR 1910.252) that hot work permits may be required to ensure such work is performed safely.

In confined space entry in particular, OSHA's concern for the safety and health of confined space personnel involved in hot work operations is well warranted. Confined spaces, by their very nature, are dangerous environments. Whenever you add hot work to the mix, the potential for additional hazards can be deadly.

This chapter includes a sample hot work program and a sample permit.

Fire watch requirements are also covered. Again, as with the sample confined space, respiratory protection, and lockout/tagout program, the hot work permitting program discussed has been used (very successfully) in the field for more than 5 years.

14.2 HOT WORK PERMIT PROGRAM (A SAMPLE)

I. INTRODUCTION

OSHA's Process Safety Management Standard (CFR 29 1910.119), Confined Space Entry Program (29 CFR 1910.146), and Respiratory Protection Standard (29 1910.134) requires organization employees, the host facility, and outside contractors to employ safe work practices when performing hot work in or near hazardous materials/chemicals or in confined spaces. Also required are hot work permits, which describe the proposed work action and the allowable work period.

II. HOT WORK DEFINITION

Hot work is defined as the use of oxy-acetylene torches, welding equipment, grinders, cutting, brazing, or similar flame-producing or spark-producing operations.

III. HOT WORK PERMIT

A hot work permit (see Figure 14.1) will be required for contractors and organization employees for any hot work performed "in or near" hazardous material/chemical processes, facilities, and confined spaces as follows:

(1) Work *on* tanks, containers, piping feed systems or ancillary equipment containing chemicals or fuels, and work in confined spaces
(2) Work within 25 feet of a digestor (direct flame only)
(3) Work in chemical rooms or on any part of non-diluted chemical system
    Note: Special precautions must be used when performing gas welding on this system: Never use acetylene or propane in the presence of chlorine.
(4) Work within 25 feet of any flammable/combustible materials with NFPA fire rating of 2 or greater
(5) Wherever confined space entry testing indicates a hazardous atmosphere

HOT WORK PERMIT
for compliance with
(OSHA 1910.119 – PROCESS SAFETY MANAGEMENT STANDARD)

SECTION 1 (Please Print)

WORK DESCRIPTION — TIME: BEGIN — END

TOOLS TO BE USED/SPECIAL HAZARDS — DATE:

PERMIT ISSUED TO (NAME/COMPANY) — PERMIT ISSUED BY (NAME)

SECTION 2

To be verified by supervisor of area where work is to be performed

ITEM	YES	NO	N/A	COMMENTS	ITEM	YES	NO	N/A	COMMENTS
Lines/Tanks Washed					Interfacing Areas Notified				
Lines/Tanks Drained					Extinguisher Present				
Lines/Tanks Pressure Vented					Confined Space				
Lines Blinded/ Disconnected					Oxygen Level*				Level:
Valves Off Locked/Tagged					L. E. L. *				Level:
Power Off Locked Tagged					Fire Watch	Name/Company:			

I certify all the items above have been completed and hereby authorize this permit.

HOST or CONTRACTOR

Atmosphere Tester Signature__________ Site Supervisor's Initials________
(Asterisk items only)

SECTION 3

To be completed by Maintenance or Contractor personnel.

ITEM	YES	NO	N/A	COMMENTS	ITEM	YES	NO	N/A	COMMENTS
Lines Blinded/ Disconnected					Glasses/Gloves				
Valves Off Locked/Tagged					Protective Cloth				
Power Off Locked Tagged					Area Roped/ Barricade/ Signs In Place				
Air Mask					Fire Watch Present				
Air Bottles Checked					Screens & Curtains				

I certify all the items above have been completed and hereby authorize this permit.

Maintenance/Contractor signature__________

White Tag copy - Display at work site — Canary copy - Supervisor's Log copy

Figure 14.1 Hot work permit (a sample).

(6) Wherever a "Hot Work Permit Required" sign is posted
(7) Work, where in the organization's supervisor's judgment, ignition/explosion of chemicals could occur from sparks, hot slag, etc.
(8) (Organization Personnel Only) Work anywhere within fenceline of Organization *interceptor* pumping stations

Note: Hot work permits *are not* required when the potential for the hazard can be removed throughout the duration of work. This can be accomplished by disconnecting and flushing lines.

(9) Work anywhere within 25 feet of an *interceptor line* excavation (no matter the depth, length, width, or other excavation dimension)

When contractors perform work on organization property and lines, organization construction project engineers and/or representatives from the Safety Division will point out to the contractors where hot work permit procedures will be required.

Hot work permits expire upon completion of each indicated task and at the end of the workday. A new permit must be completed and issued at the beginning of each workday.

After organization supervisor completes Sections 1 and 2 of the hot work permit, the white tag copy will be passed on to organization/maintenance/private contractor, who will complete Section 3 and display the white tag copy of permit at the work site until hot work operations are completed. Air sampling to determine oxygen level and LEL will be completed and readings entered on hot work permit form by either the contractor or organization, depending on who is to perform the hot work. The canary copy will remain in the hot work permit issuance book maintained by organization supervisor.

IV. PERMIT INFORMATION/SAFE WORK PRACTICES

The hot work permit lists require safe work practices that must be documented in the permit and followed during the specified hot work operations. Any of the safe work practice items listed in the permit that are not applicable to a particular work operation must be noted in the appropriate comment area.

V. FILLING OUT THE HOT WORK PERMIT

The organization hot work permit is divided into three sections. The procedure for completing each of these three sections is described as follows:

(1) *Section 1*—must be completed by the designated organization work site supervisor. This supervisor must be a minimum of Grade 17 (Grade 10 or designated interceptor department personnel) and must be trained on hot work safe work practices and permitting procedures.

(2) *Section 2*—all actions must be completed and verified by the designated organization work site supervisor. This supervisor must be a minimum Grade 17 (Grade 10 or designated interceptor department personnel) and must have been trained in hot work safe work practices and permitting procedures.

(3) *Sections 3*—must be completed by the organization maintenance person or contractor person who is to perform the hot work.

The hot work cannot proceed unless all items in Sections 2 and 3 are checked "yes" or indicated as non-applicable (N/A) in the comment section. A check of "yes" by the designated supervisor, organization maintenance person, or contractor verifies that the designated supervisor has established that the indicated safe work practice is accomplished.

A. Section 1 Items

(1) *Work Description*—Describes hot work to be accomplished, and location.

(2) *Tools to Be Used/Special Hazards*—Describes tools to be used in performing hot work, e.g., welding machine, oxy-acetylene, brazing equipment, etc. Any special hazards (such as concerns for air quality) while performing the hot work must be entered here.

(3) *Permit Issued To (Name/Company)*—Organization person or contractor person and company name (must be designated person and not just the company name) is entered here.

(4) *Time*—Enter time hot work is to begin and end.

(5) *Date*—Effective date of the permit. Note: A new permit must be issued each day.

(6) *Permit Issued By (Name)*—Organization person who issued permit.

B. Section 2 Items

(1) *Lines/Tanks Washed*—Lines/tanks have been washed and are free of chemical and flammable residue.

(2) *Lines/Tanks Drained*—Lines/tanks have been drained and flushed of chemicals/flammables.

(3) *Lines/Tanks Pressure Vented*—Pressure lines/tanks have been vented.

(4) *Lines Blinded/Disconnected*—Lines undergoing hot work that might pose hazard to other connected parts/systems have been blinded/disconnected.

(5) *Valves Off/Locked/Tagged*—Affected valves in system/lines are closed, and lockout/tagout is complete. Locks/tags should be used by both site personnel and contractor personnel if system to be locked/tagged out is a site system.

(6) *Power Off/Locked/Tagged*—All stored energy sources of affected system are placed in zero energy state and locked/tagged. Mutual locks/tags should be used on site systems.

(7) *Interfacing Areas Notified*—Adjacent buildings, rooms, compartments that might be affected by hot work must be inspected, and personnel in affected areas must be notified of adjacent hot work.

(8) *Extinguisher Present*—An appropriate rated fire extinguisher is present at the hot work site.

(9) *Confined Space*—For hot work performed in a confined space, confined space entry by permit procedures have been completed. Always ensure oxygen and LEL levels are listed.

(10) *Oxygen Level*—For hot work performed in a confined space, enter the oxygen level in the comment section. This entry is made by organization's qualified atmosphere tester for hot work performed by organization personnel and by the contractor's qualified atmosphere tester for hot work performed by the contractor.

(11) *LEL*—For hot work performed in a confined space, enter the lower explosive limit (LEL) in the comment section. This entry is made by organization's qualified atmosphere tester for hot work performed by organization personnel and by the contractor's qualified atmosphere tester for hot work performed by the contractor.

(12) *Fire Watch*—All hot work requires the stationing of a designated Fire Watch. The Fire Watch must be trained and remain on station for at least 30 minutes after completion of hot work to guard against reflash. For hot work not involving the application of direct flame, Fire Watch can be relieved of Fire Watch duties after a thorough inspection of hot work area has been made to ensure that there is no danger of reflash produced from hot metal chips/slag residue. Fire Watch's name and company name must be entered on permit.

(13) *Atmosphere Tester Signature*—If confined space LEL and oxygen testing is required, a qualified atmosphere tester for organization will test the atmosphere and sign for hot work to be performed by organization personnel. For hot work to be performed by the contractor, the contractor's qualified atmosphere tester will test the atmosphere and sign.

(14) *Site Supervisor's Initials*—The site supervisor verifies the permit information in Sections 1 and 2.

C. Section 3 Items

(1) *Lines Blinded/Disconnected*—Lines undergoing hot work that might pose a hazard to other connected systems have been blinded/disconnected.

(2) *Valves Off Locked/Tagged*—Affected valves in system/lines are

closed, and lockout/tagout is complete. Mutual locks/tags should be used on site systems.

(3) *Power Off Locked/Tagged*—All stored energy sources of the affected system are to be placed in zero energy state and locked/tagged. Mutual locks/tags should be used on site systems.

(4) *Air Mask*—For hot work performed in confined space or where possible inhalation of fumes is excessive, the appropriate OSHA/NIOSH approved Respiratory Protection Procedure must be followed.

(5) *Air Bottle Checked*—For hot work performed in confined space or where possible inhalation of fumes is excessive, the appropriate air supply bottles are used. Note: *Only Grade D breathing air is permitted—pure oxygen is prohibited!*

(6) *Glasses/Gloves*—OSHA-approved glasses and gloves are in use.

(7) *Protective Cloth*—Covering for protecting combustibles when hot work might produce slag and sparks that could cause fire.

(8) *Area Roped Off/Barricaded with Signs*—Whenever hot work is performed in an area where normal transit of personnel takes place, it must be roped off. If a normally transited area cannot be roped off to prevent unauthorized entry, then it must be barricaded with appropriate warning signs posted.

(9) *Fire Watch Present*—Verify Fire Watch is present.

(10) *Screens and Curtains*—Welding operations must be conducted with proper screens or curtains to protect passing personnel from ultraviolet radiation.

(11) *Organization Maintenance Person/Contractor Signature*—Organization maintenance person or contractor person verifies that all indicated items have been completed.

14.3 FIRE WATCH REQUIREMENTS

As stated earlier and as shown on the hot work permit (see Figure 14.1), a Fire Watch must be assigned whenever hot work operations are being performed around hazardous materials, in confined spaces, and other times when there is the danger of fire and/or explosion from such work. OSHA has specific requirements regarding Fire Watch duties.

Fire watchers shall be required whenever welding or cutting is performed in locations where other than a minor fire might develop or any of the following conditions exist:

(1) Appreciable combustible material, in building construction or contents, closer than 35 feet (10.7 m) to the point of operation.

(2) Appreciable combustibles are more than 35 feet (10.7 m) away but are easily ignited by sparks.
(3) Wall or floor openings within a 35-foot (10.7 m) radius expose combustible material in adjacent areas, including concealed spaces in walls or floors.
(4) Combustible materials are adjacent to the opposite side of metal partitions, walls, ceilings, or roofs and are likely to be ignited by conduction or radiation.

Fire watchers shall have fire extinguishing equipment readily available and be trained in its use. They shall be familiar with facilities for sounding an alarm in the event of a fire. They shall watch for fires in all exposed areas, try to extinguish them only when obviously within the capacity of the equipment available, or otherwise sound the alarm. A fire watch shall be maintained for at least 30 minutes after completion of welding or cutting operations to detect and extinguish possible smoldering fires.

14.4 REFERENCE

OSHA. *Occupational Safety and Health Standards for General Industry, 29 CFR Part 1910,* with amendments. Washington, D.C., U.S. Department of Labor, 1995.

Afterword

CONFINED space entry always carries with it the potential for danger, from minor incident to full-blown catastrophe. These inherent problems come with the territory, a part of the spaces we must enter or ask workers to enter, and of the materials we handle or come into contact with, in our jobs and on our work sites.

OSHA's goal in setting these standards and our goal in meeting them are the same—to eliminate or reduce the hazards workers face in confined spaces and the risks they must often take on the job.

While accident and incident, injury, and fatality statistics are useful in benchmarking our progress toward safer workers and work sites, statistics sometimes allow us to lose site of the reasons safety and health regulations are important—each one of those statistics might have been, could have been, a partner, a co-worker, a parent, a child, a friend. Anyone who has narrowly escaped injury or death has said, "I'm lucky that wasn't me." OSHA's Confined Space Entry Standards are successfully working to eliminate the "luck" element—to replace it with the protection afforded by training, planning, preparation, skill, and thought.

Which would you rather trust?

Index